Latest Advancements in Micro Nano Molding Technologies – Process Developments and Optimization, Materials, Applications, Key Enabling Technologies

Latest Advancements in Micro Nano Molding Technologies – Process Developments and Optimization, Materials, Applications, Key Enabling Technologies

Editor

Guido Tosello

MDPI • Basel • Beijing • Wuhan • Barcelona • Belgrade • Manchester • Tokyo • Cluj • Tianjin

MDPI

Editor
Guido Tosello
Department of Civil and
Mechanical Engineering
Technical University of
Denmark
Kgs. Lyngby
Denmark

Editorial Office
MDPI
St. Alban-Anlage 66
4052 Basel, Switzerland

This is a reprint of articles from the Special Issue published online in the open access journal *Micromachines* (ISSN 2072-666X) (available at: www.mdpi.com/journal/micromachines/special_issues/Micro_Nano_Molding).

For citation purposes, cite each article independently as indicated on the article page online and as indicated below:

LastName, A.A.; LastName, B.B.; LastName, C.C. Article Title. *Journal Name* **Year**, *Volume Number*, Page Range.

ISBN 978-3-0365-5434-1 (Hbk)
ISBN 978-3-0365-5433-4 (PDF)

Cover image courtesy of Komeil Saeedabadi

Contents

About the Editor

Guido Tosello

Guido Tosello, PhD, is Associate Professor at the Technical University of Denmark, Department of Civl and Mechanical Engineering, Section of Manufacturing Engineering. Tosello's principal research interests are the analysis, characterization, monitoring, control, optimization, and simulation of precision molding processes at micro/nanoscales of thermoplastic materials. Technologies supporting precision/micro/nano molding processes are of research interest: advanced process chain for micro/nano tools manufacturing, precision and micro additive manufacturing, and dimensional and surface micro/nano metrology. Guido Tosello is the recipient of the Technical University of Denmark Best PhD Research Work 2008 Prize for his PhD thesis "Precision Moulding of Polymer Micro Components"; of the 2012 Alan Glanvill Award by The Institute of Materials, Minerals, and Mining (IOM3) (UK), given as recognition for research of merit in the field of polymeric materials; of the Young Research Award 2014 from the Polymer Processing Society (USA) in recognition of scientific achievements and research excellence in polymer processing within 6 years from PhD graduation; and of the Outstanding Reviewer Award 2016 of the Institute of Physics (UK) for his contribution to the Journal of Microengineering and Micromechanics. Guido Tosello has been Project Coordinator of the Horizon 2020 European Marie Skłodowska-Curie Innovative Training Networks MICROMAN "Process Fingerprint for Zero-defect Net-shape MICROMANufacturing"(2015–2019) (http://www.microman.mek.dtu.dk/), is Coordinator of the DIGIMAN4.0 "DIGItal MANufacturing Technologies for Zero-defect Industry 4.0 Production"(2019-2024) (http://www.digiman4-0.mek.dtu.dk/), and is Editor of the book 'Micro Injection Molding' published by Carl Hanser Verlag in 2018. He holds an Executive MBA degree from the DTU Executive School of Business.

micromachines

MDPI

Editorial

Latest Advancements in Micro Nano Molding Technologies—Process Developments and Optimization, Materials, Applications, Key Enabling Technologies

Guido Tosello

Department of Mechanical Engineering, Technical University of Denmark, Produktionstorvet, Building 427A, 2800 Kongens Lyngby, Denmark; guto@mek.dtu.dk; Tel.: +45-45-25-4893

Citation: Tosello, G. Latest Advancements in Micro Nano Molding Technologies—Process Developments and Optimization, Materials, Applications, Key Enabling Technologies. *Micromachines* **2022**, *13*, 609. https://doi.org/10.3390/mi13040609

Received: 6 April 2022
Accepted: 7 April 2022
Published: 13 April 2022

Publisher's Note: MDPI stays neutral with regard to jurisdictional claims in published maps and institutional affiliations.

Micro and nano molding technologies are continuously being developed due to enduring trends such as increasing miniaturization and the higher functional integration of products, devices and systems. Furthermore, with the introduction of high-engineering-performance polymers, feedstocks and composites, new opportunities in terms of materials properties can be exploited, and consequently, more micro product and micro/nano structured surfaces are currently being designed and manufactured.

Innovations in micro and nano molding techniques are seen in different processes employed in production (e.g., injection molding, micro injection molding, powder micro molding, two-component molding, compression molding, hot embossing, nanoimprint lithography); in the use of new and functional materials including, e.g., nanocomposites; for an ever-increasing number of applications (health-care devices, micro implants, micro analytics systems, mobility and communications products, optical elements, micro electromechanical systems, sensors, micro molded interconnected devices, etc.); in several key enabling technologies that support the successful realization of micro and nano molding processes (micro and nano tooling technologies, process monitoring techniques, micro and nano metrology methods for quality control, simulation, rapid prototyping techniques for micro product development, etc.) and their integration into new manufacturing process chains.

Accordingly, this Special Issue seeks to showcase research papers focusing on the latest developments in micro and nano scale manufacturing using molding techniques as well as their related key enabling technologies to produce both micro products and micro/nano structured surfaces.

The Special Issues consists of 10 original research papers and 2 review papers, which cover fundamental molding process technology development, key enabling technologies, as well as the design and application of these technologies for the fabrication of micro/nano devices and micro structured components.

The papers included in the Special Issue address research, development and recent advancements in four main domains of micro/nano molding: (1) process technology developments and characterization; (2) modeling and simulation; (3) tooling technologies and micro tool design; (4) applications.

(1) Process technology developments and characterization. Calaon et al. [1] analyzed and compared the process capability and design of conventional injection molding and of micro injection molding machines; Wöhner et al. [2] characterized the formation of blisters in film micro insert molding; Loaldi et al. [3] integrated direct ink writing with injection molding to generate micro conductive tracks in polymer devices.

(2) Modeling and simulation. Weng et al. [4] modeled the demolding phase in nano scale molding by using molecular dynamic simulations; Loaldi et al. [5] simulated the micro injection molding process of both three-dimensional micro parts and mi-

cro structured components by applying multi-scale meshing and virtual design of experiment techniques.

(3) Tooling technologies and micro tool design. Wang et al. [6] developed the air-shielding electrochemical micromachining (AS-EMM) process to improve the generation of microstructures on metal surfaces; Li et al. [7] analyzed the effects of machining errors on the optical performance of optical aspheric components in ultra-precision diamond turning; Tucker et al. [8] characterized different venting design solutions in micro molding tools and the influence of micro injection molding process parameters on air traps and adiabatic heating.

(4) Applications. Kim et al. [9] presented the production of a miniaturized out-of-plane compliant bistable mechanism (OBM) via micro injection molding; Wu et al. [10] reviewed the state of the art and perspectives on silicon waveguide crossings and discussed the use of polymers as vertical directional coupler materials; Cheng et al. [11] reviewed the design and emerging trends of grating couplers on silicon photonics and presented the possibility of a polymer-based micro packaging application of plasmonic surfaces; Takehara et al. [12] fabricated microneedle arrays in Poly (L-lactic Acid) via micro hot embossing/compression molding and studied the effect of the thermal history on the material crystallinity during the process.

We wish to thank all of the authors who submitted their papers to this Special Issue, entitled "Latest Advancements in Micro Nano Molding Technologies—Process Developments and Optimization, Materials, Applications, Key Enabling Technologies". We would also like to acknowledge all of the reviewers whose careful and timely reviews ensured the high quality of this Special Issue.

The support and funding from the European Commission Horizon2020 Framework Programme for Research and Innovation through the ProSurf project ("High Precision Process Chains for the Mass Production of Functional Structured Surfaces", http://www.prosurf-project.eu/, accessed on 5 April 2022, 2018–2021, Project ID: 767589), as well as the Marie Skłodowska-Curie Action Innovative Training Networks MICROMAN ("Process Fingerprint for Zero-defect Net-shape MICROMANufacturing", http://www.microman.mek.dtu.dk/, accessed on 5 April 2022, 2015–2019, Project ID: 674801) and DIGIMAN4.0 ("DIGItal MANufacturing Technologies for Zero-defect Industry 4.0 Production", http://www.digiman4-0.mek.dtu, accessed on 5 April 2022, 2019–2024, Project ID: 814225), is acknowledged. The support and funding from the Danish Innovation Fund (https://innovationsfonden.dk/en, accessed on 5 April 2022), through the research projects MADE DIGITAL, Manufacturing Academy of Denmark (http://en.made.dk/, accessed on 5 April 2022, 2017–2020, Project ID: 6151-00006B), Work Package WP3 "Digital manufacturing processes" and QRprod ("QR coding in high-speed production of plastic products and medical tablets", 2016–2019 Project ID: 5163-00001B) is acknowledged.

Conflicts of Interest: The author declares no conflict of interest.

References

1. Calaon, M.; Baruffi, F.; Fantoni, G.; Cirri, I.; Santochi, M.; Hansen, H.; Tosello, G. Functional Analysis Validation of Micro and Conventional Injection Molding Machines Performances Based on Process Precision and Accuracy for Micro Manufacturing. *Micromachines* **2020**, *11*, 1115. [CrossRef] [PubMed]
2. Wöhner, T.; Islam, A.; Hansen, H.; Tosello, G.; Whiteside, B. Blister Formation in Film Insert Moulding. *Micromachines* **2020**, *11*, 424. [CrossRef] [PubMed]
3. Loaldi, D.; Piccolo, L.; Brown, E.; Tosello, G.; Shemelya, C.; Masato, D. Hybrid Process Chain for the Integration of Direct Ink Writing and Polymer Injection Molding. *Micromachines* **2020**, *11*, 509. [CrossRef]
4. Weng, C.; Yang, D.; Zhou, M. Molecular Dynamics Simulations on the Demolding Process for Nanostructures with Different Aspect Ratios in Injection Molding. *Micromachines* **2019**, *10*, 636. [CrossRef] [PubMed]
5. Loaldi, D.; Regi, F.; Baruffi, F.; Calaon, M.; Quagliotti, D.; Zhang, Y.; Tosello, G. Experimental Validation of Injection Molding Simulations of 3D Microparts and Microstructured Components Using Virtual Design of Experiments and Multi-Scale Modeling. *Micromachines* **2020**, *11*, 614. [CrossRef] [PubMed]

6.   Wang, M.; Shang, Y.; He, K.; Xu, X.; Chen, G. Optimization of Nozzle Inclination and Process Parameters in Air-Shielding Electrochemical Micromachining. *Micromachines* **2019**, *10*, 846. [CrossRef]
7.   Li, Y.; Zhang, Y.; Lin, J.; Yi, A.; Zhou, X. Effects of Machining Errors on Optical Performance of Optical Aspheric Components in Ultra-Precision Diamond Turning. *Micromachines* **2020**, *11*, 331. [CrossRef] [PubMed]
8.   Tucker, M.; Griffiths, C.; Rees, A.; Llewelyn, G. High Temperature Adiabatic Heating in μ-IM Mould Cavities—A Case for Venting Design Solutions. *Micromachines* **2020**, *11*, 358. [CrossRef] [PubMed]
9.   Kim, W.; Han, S. Microinjection Molding of Out-of-Plane Bistable Mechanisms. *Micromachines* **2020**, *11*, 155. [CrossRef] [PubMed]
10.   Wu, S.; Mu, X.; Cheng, L.; Mao, S.; Fu, H. State-of-the-Art and Perspectives on Silicon Waveguide Crossings: A Review. *Micromachines* **2020**, *11*, 326. [CrossRef] [PubMed]
11.   Cheng, L.; Mao, S.; Li, Z.; Han, Y.; Fu, H. Grating Couplers on Silicon Photonics: Design Principles, Emerging Trends and Practical Issues. *Micromachines* **2020**, *11*, 666. [CrossRef] [PubMed]
12.   Takehara, H.; Hadano, Y.; Kanda, Y.; Ichiki, T. Effect of the Thermal History on the Crystallinity of Poly (L-lactic Acid) During the Micromolding Process. *Micromachines* **2020**, *11*, 452. [CrossRef] [PubMed]

micromachines

MDPI

Article

Blister Formation in Film Insert Moulding

Timo Wöhner [1], Aminul Islam [1,*] , Hans N. Hansen [1] , Guido Tosello [1] and
Ben R. Whiteside [2]

[1] Department of Mechanical Engineering, Technical University of Denmark, 2800 Kgs. Lyngby, Denmark;
    timo_woehner@hotmail.de (T.W.); hnha@mek.dtu.dk (H.N.H.); guto@mek.dtu.dk (G.T.)
[2] The Centre for Polymer Micro and Nano Technology (Polymer MNT), University of Bradford,
    BD7 1DP Bradford, UK; b.r.whiteside@bradford.ac.uk
*   Correspondence: mais@mek.dtu.dk; Tel.: +45-452-548-96

Received: 19 March 2020; Accepted: 16 April 2020; Published: 17 April 2020

Abstract: The formation of blister in the injection moulded parts, especially in the film insert moulded parts, is one of most significant causes of part rejection due to cosmetic requirements or functionality issues. The mechanism and physics of blister formation for molded parts are not well-understood by the state-of-the-art literature. The current paper increases the fundamental understanding of the causes for blister formation. In the experiment, a membrane strip of 5 mm in width was overmoulded with Polypropylene (PP), which formed a disc-shaped part with a diameter of 17.25 mm and a thickness of 500 μm. To investigate the influence of the processing parameters, a full factorial design of experiments (DoE) setup was conducted, including mould temperature (T_m), barrel temperature (T_b), injection speed (V_i) and packing pressure (P_p) as variables. The degree of blistering at the surface was characterized by the areal surface roughness parameters Spk and Smr1, measured with a confocal laser microscope. The measurements were taken on the 10 mm long section of the membrane surface in the centre of the moulded part across the entire width of the film. In addition, the film insert moulding (FIM)-process was simulated and the average shrinkage of the substrate material under the membrane was investigated. Eventually, a method and processing window could be defined that could produce blister-free parts.

Keywords: film insert moulding; surface analysis; blister formation; simulation

1. Introduction

In film insert moulding (FIM) a preformed film is inserted in the mould and subsequently overmoulded. This injection moulding technique is mainly used to create decorative plastic parts with a high-quality surface finish and enables one-step fabrication of plastic components with a decorated or functional surface. FIM parts can be found for example in automotive industry, medical technology, design and fashion products, household appliances, as well as for mobile phones and many other applications [1]. Apart from this, FIM has been used to add functional features like nano-patterned structures [2] or Radio Frequency Identification tags (RFID-tags) [3] to the surface of injection moulded parts. Furthermore, the use of FIM to overcome the mechanical problems of weld lines in glass fibre filled thermoplastics has been reported [4]. Despite its manifold applications, the number of scientific publications on the topic is quite low and many fundamental issues need to be addressed for the realization of defect free FIM and to exploit the full benefit of FIM.

One of the key areas of research on FIM is the effect of the change in the cooling rate of the part when a film is inserted. A plastic film insert decreases the heat conduction between melt and mould by acting as a barrier layer, which influences the filling and cooling behaviour of the part. The warpage resulting from these uneven cooling conditions was investigated in some publications such as [1] and [5–7]. Chen et al. [7] reported an asymmetric flow behaviour of Polypropylene when a film was inserted, where the mould side covered by the film showed a flow leading effect. Furthermore, they found that the film insert slows down the increase of the mould surface temperature under the film insert and reduces the maximum mould surface temperature reached under the film insert compared to a conventional moulding process with the same mould. The warpage caused by these asymmetric cooling conditions increases with an increase in film thickness and melt temperature and decreases with an increase in mould temperature. These results are in good agreement with those found by Kim, et al. [8] who investigated the increase in the asymmetry of the temperature distribution with increasing film thickness.

The warpage of FIM-parts can be reduced or even reversed by an annealing step when an unannealed film was used. According to Kim, et al. [6] this effect is caused by the relaxation of residual stresses from film production, which can lead to a negative coefficient of thermal expansion (CTE) of the inserted film. In semi-crystalline materials, the formation of an amorphous skin layer can be suppressed due to the reduced cooling rate in the sections covered by the film. Kim et al. [9] found a higher degree of crystallization on the side of the part covered by the film, while the uncovered side showed an amorphous skin layer. The same effect was found by Chen et al. [7] who described an increase in crystallinity and the size of the crystallites with a decrease in the cooling rate due to the film insert, which also leads to an increase in warpage. Warpage in FIM has been investigated in several pieces of research and is considered as one of the main bottlenecks associated with the FIM process [10–13].

Another area of interest in FIM research deals with the adhesion between the film insert and the injection moulded substrate. Leong et al. [14] stated that the bonding strength can be improved by increasing injection speed and packing pressure. An increase in the barrel temperature below a certain limit increases the bonding strength, while the effect of an increase in barrel temperature above this level was found to be negligible. In [15], they describe a reduced peel strength for very high injection speeds, which is explained by a reduced entanglement of polymer chains across the interface between film insert and substrate due to an increase in molecular orientation. An additional annealing step led to even lower peel strength due to an increase in crystallization and the resulting reduction in loose molecules which are available for entanglements across the interface.

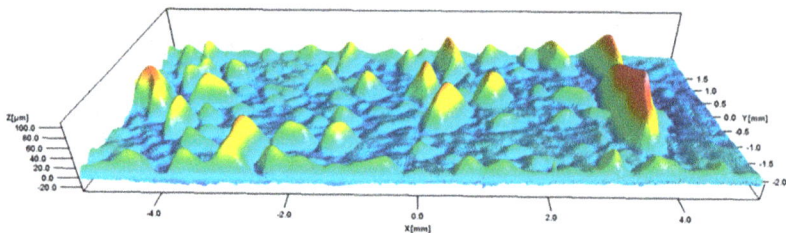

Figure 1. Blistered surface of the film insert after moulding (Z-axis scale 10:1). Moulding parameters: Mould temperature: 40 °C, barrel temperature: 200 °C.

In the current work the combination of a dual layer membrane made from a thin membrane layer, a non-woven support layer and polypropylene (PP) as an overmoulding material were tested for FIM. The ultimate goal of the work was to use the materials and process for the production of moulding-based methanol fuel cell (DMFC) containers, which are supposed to be used in hearing aid applications [16]. During this investigation, it was found that, for some combinations of the process, parameter blisters occurred in the membrane layer (see Figure 1). This phenomenon, to the author's knowledge, has not been described in the literature before. For the investigation of the blisters, a confocal laser microscopy and film insert moulding simulations have been used. A full factorial design of experiments (DoE) was performed to define a process window that allows for blister free moulding and to investigate the relationship between the average shrinkage under the membrane found by means of simulation and the blistered surface analysis.

2. Materials and Methods

2.1. Materials

In this work, a SABEU TRAKETCH®-membrane was used as film insert. This membrane has an overall thickness of 200 ± 20 µm. It consists of a 23 µm thick, porous PET-layer which has an oleophobic coating and a non-woven PP support. The pore size is 0.22 ± 0.02 µm and the pore-density is 270×10^6 pores / cm^2. PP was used as substrate material (PP579S from SABIC, Riyadh, Saudi Arabia). This polypropylene grade shows a high melt flow index, provides high stiffness and low warpage tendency according to the supplier. Further background information on the material selection can be found in [17].

2.2. Part Design

The moulded part for this investigation was disc-shaped with a diameter of 17.25 mm and a thickness of 500 µm. In the centre of the disc, a 5 mm wide strip of the film insert was overmoulded which covered the whole diameter of the disc (see Figure 2). The film insert was aligned in the mould by placing it into a recess in the movable mould half, so that the support layer was overmoulded. It was fixed using double-sided adhesive tape at both ends of the film insert and additionally clamped between the mould halves.

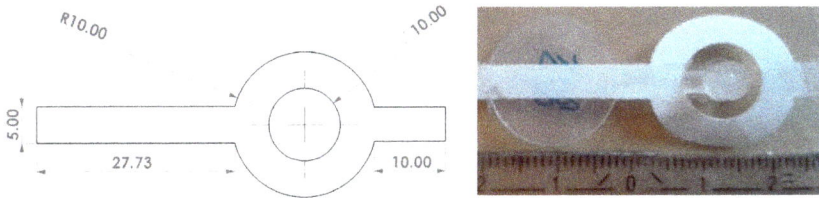

Figure 2. Design of the film insert blank (**left**, all dimensions in mm), film insert moulding (FIM)-part including sprue, runner, gate and untrimmed film insert (**right**).

2.3. Moulding Experiments

For the moulding experiments, a Wittmann–Battenfeld Micro Power 15 micro injection moulding machine was employed (see Figure 3, which also shows the mould insert used for the FIM experiment). This machine uses a screw (diameter 14 mm) for plastification of the material and an injection plunger with a diameter of 5 mm for injecting the material. A full factorial DoE analysis was used to investigate the influence of the injection moulding parameters like mould temperature (T_m), barrel temperature (T_b), injection speed (V_i) and packing pressure (P_p). The chosen parameter levels can be found in Table 1.

Table 1. Machine-settings for the moulding experiments.

Parameter	Low Level	Medium Level	High Level
T_m	40 °C	-	60 °C
T_b	200 °C	235 °C	270 °C
V_i	100 mm/s	200 mm/s	300 mm/s
P_p	72 bar	99 bar	126 bar

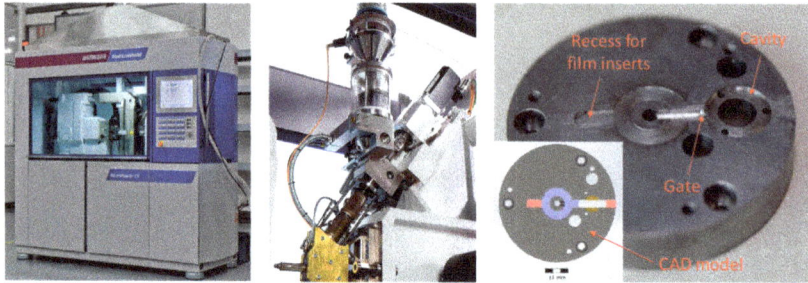

Figure 3. Wittmann–Battenfeld Micro Power 15 micro injection moulding machine used for the experiment (**left**), the injection unit of the of moulding machine (**middle**); and mould insert used in the tool to injection moulding of the FIM part (**right**).

The barrel temperatures were chosen to cover the temperature ranges given in the processing datasheet for the used polypropylene. The entire mould temperature range suggested in the datasheet could not be implemented, because the mould temperature control was based on electrical heating cartridges. This set-up did not allow for temperatures below room temperature. A temperature in the middle of the suggested T_m range was therefore used as low-level. A mould temperature of 60 °C is recommended in the datasheet for thick walled parts, but is used in these experiments to cover the range of recommended values as well as possible. To reduce the number of experimental runs, only two levels for the mould temperature were chosen. This decision was based on a paper by Chen et al. [7]. They found that the influence of mould temperature on warpage, crystallinity and crystal size shows a linear behaviour using PP in FIM. Therefore, a linear influence of T_m is expected in these experiments.

The packing pressure was set to 40%, 55% and 70% of the Maximum Injection Pressure (MIP). To define the MIP the DoE-runs showing the highest injection pressure were identified by means of simulations. The according processing parameters were used in the machine and the MIP was read from the machine as 180 bar. The packing time was set to four seconds throughout the set of experiments. This packing time is long enough to guarantee for a frozen gate. The maximum time for the gate freeze-off found in the simulations was below three seconds. Therefore, the packing time will have no influence on the experimental results. An additional cooling time of three seconds was used, to assure that the entire part is solidified. The influence of these two parameters can, therefore, be excluded. For each of the 54 runs of the DoE-table, five parts with membrane were moulded. Due to the manual insertion of the film insert, a long mould open time (30 s) had to be chosen. To stabilize the process conditions, five parts were moulded without membrane and rejected before a part with film insert was moulded.

2.4. Simulations

Computer modeling and simulation is revolutionizing manufacturing industries, allowing for optimizing product development process, improving resource management and enhancing the product quality [18,19]. Simulation aids the entire process, starting from component design, mold design, and manufacturing, material selection, molding conditions' enhancement and ending with cooling optimization (shrinkage and warpage analysis). At present, it is possible to solve 3D problems by FEM including crystallization models to account for the changes in morphology that occur when semi-crystalline thermoplastics solidify [20,21]. For the current investigation, the "in-mould label overmoulding"-feature of Autodesk Mouldflow Simulation Insight 2016 was used to simulate the process sequence which contained filling, packing and cooling. This feature takes the influence of the film insert on flow and cooling behaviour into consideration [22]. The film insert was modelled as a 200 μm thick PP-film. The simulation model furthermore contained the mould, venting locations and the heating cartridges. A multi-scale mesh was used to achieve a sufficient mesh resolution at the edge of the film insert at acceptable computation times. It is assumed that the occurrence of the blisters is related to the shrinkage of the substrate material. If the shrinkage in the substrate is higher than the contraction of the film insert during the cooling phase of the process, a delamination between the film insert and the substrate material can occur [23]. For that reason, the average volumetric shrinkage of the substrate under the film insert was investigated in the simulations. Therefore, this shrinkage value was taken at 16 different locations under the film insert on the simulated part and the average of these values was used for the evaluation of the DoE.

2.5. Metrology

To characterize the degree of blistering of the membrane surface an area of 5×10 mm^2 in the centre of the disk-shaped part was measured using the 10× magnification lens of an OLYMPUS LEXT 4100 confocal laser scanning microscope (Olympus, Tokyo, Japan) in stitching mode in combination with post processing with SPIP™ image metrology software by Image Metrology A/S (version 6.4.3). This post processing routine contained steps for tilt compensation, warpage compensation and the application of a median filter to remove speckle noise. Eventually, an area of interest was defined (see Figure 4), the zero level of the measurement was set to the most frequent height level and the areal surface roughness parameters were calculated. To investigate the influence of the process parameters on the emergence of blisters the areal surface roughness parameters Spk (reduced peak height) and Smr1 (peak material portion) were evaluated. Spk, the reduced peak height, is a measure for the average height of the protruding peaks above the core roughness. The averaging process due to the calculation of this parameter reduces the influence of outliers. Smr1 is the material ratio above the core roughness and therefore a measure for the area covered by blisters.

Figure 4. Olympus LEXT OLS 4000 3D measuring laser microscope used for the analysis (**left**), Measurement area of 5×10 mm^2 on the membrane in the center of the part—red area (**right**).

3. Results and Analysis

In the following section, the results of the DoE are presented. The points in the graphs indicate the mean value of all observations obtained at the indicated level of the parameter. Error bars are used to indicate a 95% confidence interval based on the estimated standard error of the mean. For each experiment, 5 parts were molded, evaluated and average results are presented in the paper, hence the error bars indicate the standard deviation calculated from 5 data points.

3.1. Smr1

According to the evaluation of the DoE, T_b had the greatest influence on the Smr1 value. Increasing T_b leads to a decrease in the portion of the surface, which is covered by blisters. In this evaluation, a 2nd order regression model was used to identify the behaviour for the parameters with three levels. The regression model indicated a non-linear dependency between T_b and Smr1 presented in Figure 5. Furthermore, it was found that, T_m and P_p had a statistically significant influence on

Smr1. For both, an increase in the input lead to a decreasing Smr1 value. Only V_i was found to have a non-significant effect at the chosen statistical significance level ($\alpha = 0.05$). The only significant two-factor interaction found was between T_m and T_b (see Figure 6).

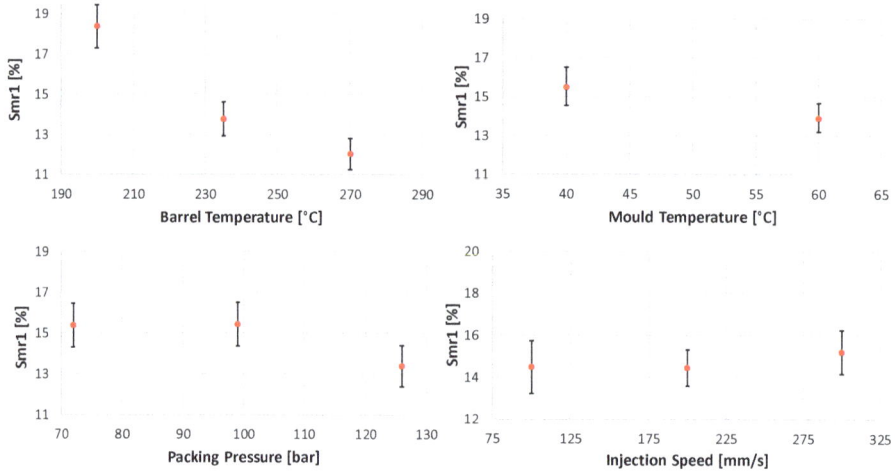

Figure 5. Main Effect of chosen parameters on the roughness parameter Smr1: effects of barrel temperature—T_b on Smr1 (**top left**); effects of mould temperature—T_m on Smr1 (**top right**); effects of packing pressure—P_p on Smr1 (**bottom left**); and effects of injection speed—V_i on Smr1 (**bottom right**).

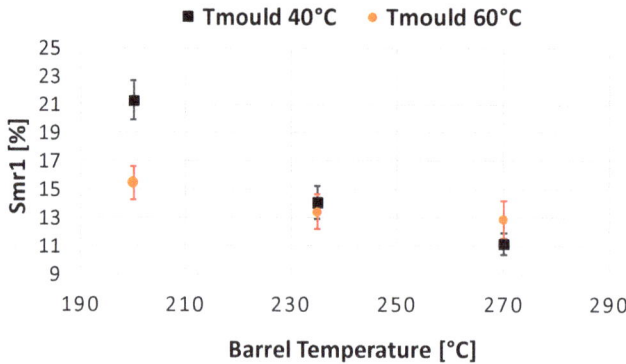

Figure 6. Two factor interaction of mould temperature (T_m) and barrel temperature (T_b) on Smr1.

3.2. Spk

The trends in the main effect plots are similar to those found for Smr1 (see Figure 7) This was expectable since higher blisters also lead to a larger share of the surface above the core roughness. According to the analysis of variance (ANOVA) for $\alpha=0.05$, the only nonsignificant main effect is, as was found for Smr1, V_i. In contrast to Smr1, Spk is influenced by three two-factor interactions. These are the interactions between T_b and T_m, T_b and P_p and V_i and T_m (see Figure 8).

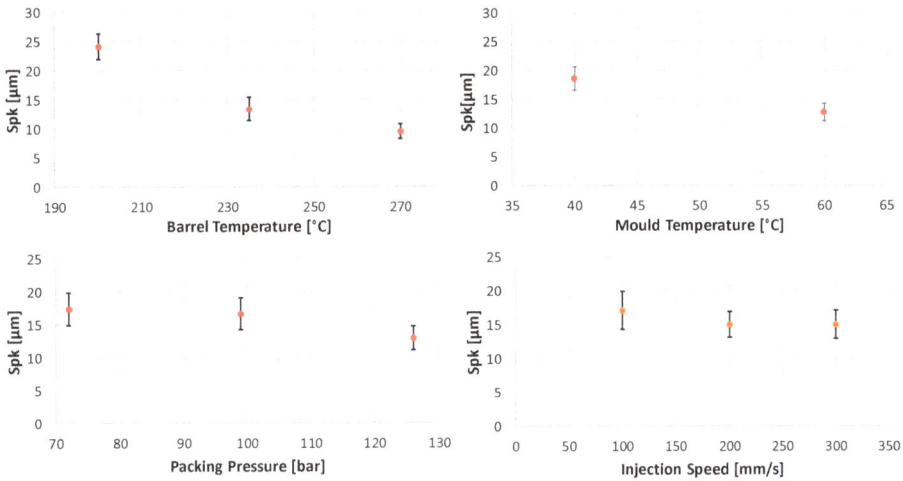

Figure 7. Main Effect of chosen parameters on the roughness parameter Spk: main effects of barrel temperature—T_b on Spk (**top left**); main effects of mould temperature—T_m on Spk (**top right**); main effects of packing pressure—P_p on Spk (**bottom left**); and main effects of injection speed—V_i on Spk (**bottom right**).

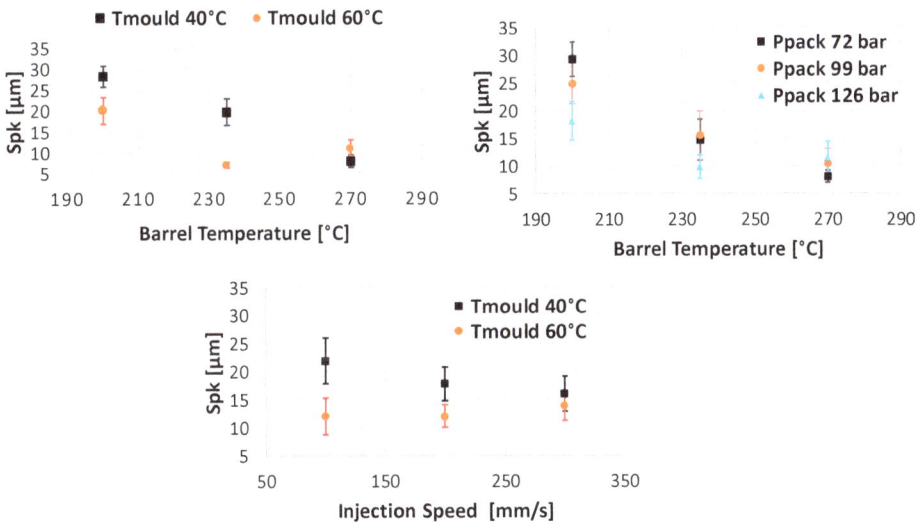

Figure 8. Two factor interactions for Spk: barrel temperature —T_b and mould temperature—T_m (**top left**); barrel temperature—T_b and packing pressure—P_p (**top right**); injection speed—V_i and mould temperature—T_m (**bottom**).

3.3. Average Volumetric Shrinkage under the Film Insert

The evaluation of the simulation showed another order of significance of the process parameters (see Figure 9). The highest influence on this result was from P_p, followed by T_m. T_b and V_i did not show any significant influence. Unlike the results for the evaluation of Smr1 and Spk, all the parameters showed a linear influence in the simulation. A significant two-factor interaction was found between T_b and P_p (see Figure 10). Main effect and interactions plots obtained by simulation contain error bars in the following pictures. The evaluation is done based on an average value of a path in the center of the part, as shown in the left-hand side picture of Figure 10. The error bars show how much the values of the "Average Volumetric Shrinkage" are spread along this path. Since the data are based on the 54 runs of the "virtual" DoE, the values at the different points in the graphs show a certain distribution.

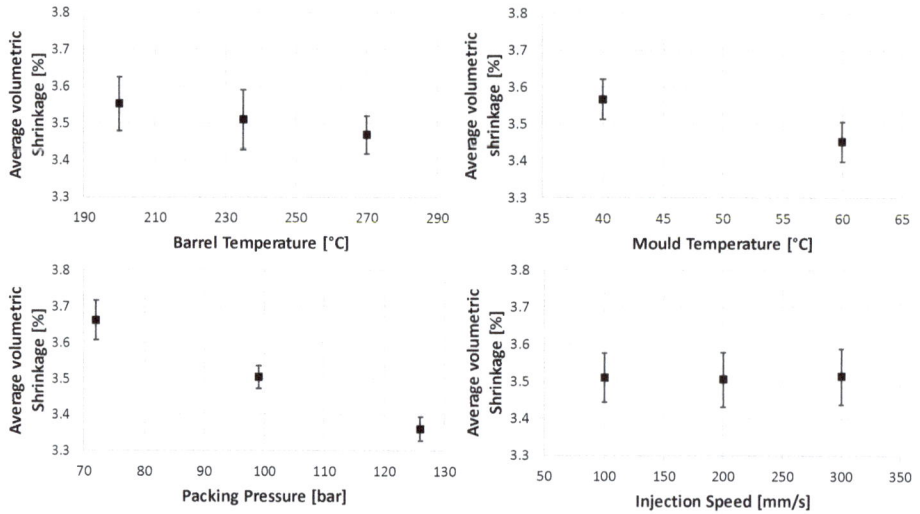

Figure 9. Main effects of chosen parameters on Average Volumetric Shrinkage (AVS): main effects of barrel temperature—T_b on AVS (**top left**); main effects of mould temperature—T_m on AVS (**top right**); main effects of packing pressure—P_p on AVS (**bottom left**); and main effects of injection speed—V_i on AVS (**bottom right**).

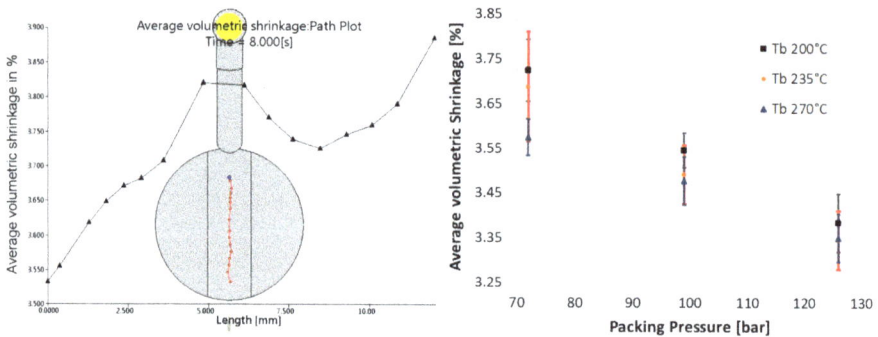

Figure 10. Average volumetric shrinkage path plot along the middle of the part (**left**), Two-factor interaction of T_b and P_p for "Average volumetric shrinkage" (**right**).

4. Discussion

According to the experimental results, the blister formation mainly depends on the process temperatures. A higher T_b mainly reduces the viscosity of the injected substrate material and increases the cooling time, while an increase in T_m increases the cooling time and reduces the thickness of the frozen layer at the interface between mould and melt and between melt and the film insert. This can lead to a more homogenous filling behaviour of the gaps between the fibres of the non-woven support layer, therefore, to a more homogenous cooling and shrinking behaviour under the membrane, which then would lead to a reduction in the blisters.

In addition, higher temperatures lead to a longer effective packing phase, since the gate freezes off later. This extended packing phase could reduce the overall shrinkage of the part, and the emergence of blisters. Furthermore, this effect could be a possible explanation for the interaction between P_p and T_b found for Spk. This interaction could indicate that the packing result cannot be further improved by increasing P_p at the highest level of T_b because, at the extended effective packing time at the highest temperature level, even the lower levels of P_p are sufficient.

The non-linearity found for T_b could be an effect of crystallization. At the prolonged cooling time at high temperatures, crystal growth can lead to an increased shrinkage, and to an increase in blister size and amount. This could also explain the interaction between T_m and T_b found for both Spk and Smr1. Another possible explanation for the non-linearity of T_b could be a stress-relaxation of the fibres of the non-woven support layer. When the melt is hot enough to raise the temperature of these fibers above the relaxation temperature, the relaxation of the frozen-in stresses of the fibres could cause their contraction. This could then lead to a delamination of the membrane from the support layer.

Eventually, a process window for blister-free parts could be defined. Parts moulded at the high level of T_m and P_p and the medium or the high level of T_b could be considered blister-free (see Figure 11) Comparing the experimental results with the simulation, a direct relation between the simulated parameter "Average Volumetric Shrinkage" and the experimental results could not be found. This is, however, no proof that the emergence of blisters is unrelated to the shrinkage of the injected substrate material under the film insert. Because of the different order of influence of the single parameters, the machine parameters recorded during the experiments were revised. It turned out that the switch-over point between the filling and packing phase was not hit properly at some of the runs. This late switch-over led to a significant increase in the maximum injection pressure and was found more often in runs with high levels of T_b and T_m. The increase in the maximum injection pressure can lead to a kind of packing which then covers the effect of the actual packing phase. This can be a reason for the lower significance of the packing phase in the experimental results.

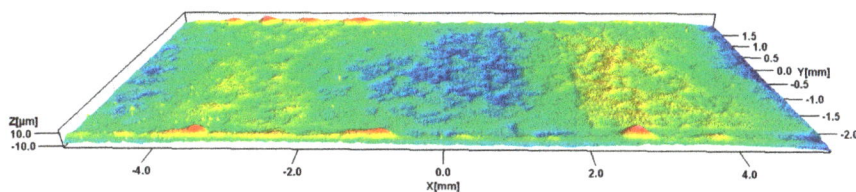

Figure 11. Part without blisters on the film insert surface (Z-axis scale 10:1) Molding parameters: T_m: 60 °C, T_b: 235 °C, P_p: 126 bar, V_i: 300 mm/s.

Further reasons for a poor reproduction of the experimental results by the simulation could be caused by the simplified model of the film insert, which did not consider the structure of the non-woven support layer, and therefore can lead to deviations from the experiments in the simulated filling behaviour and the simulated heat conduction. On top of that, there is no existing model for crystallinity for the used material. Even though a model for the calculation of crystallization effects based on [24] is implemented in the software, only few materials in the database provided the necessary input for the model when the study was conducted [25]. Therefore, the effects of crystallinity on the shrinkage are not considered in the simulations, but this is an important issue to achieve accurate simulation results [26–28]. Both these simplifications could lead to the linear behaviour found in the simulation, while T_b showed a non-linear influence in the experiments.

5. Conclusions and Outlook

In this paper, the blister formation process in FIM-parts was investigated. For that purpose, areal surface roughness parameters and simulations were used. It was found that the main influences on the formation of blisters were coming from the processing temperatures. An increase in T_m or T_b leads to a reduced blister height and a lower area portion covered by blisters. Eventually a process window for blister-free moulding could be defined for the material combinations used in the experiment. A direct relationship between the simulated shrinkage of the substrate material under the film insert could not be found, mainly due to problems with the switch-over point between the filling and packing phase. To improve the accuracy of the simulations, the experiments could be repeated with a material for which the switch-over point is easier to control. For the used material, the necessary back-pressure during the metering phase was in the range of the maximum back-pressure the machine could deliver. This is believed to be the reason for the problems with the switch-over point. Eliminating the switch-over problems would lead to a better comparability of simulations and experiments. Furthermore, an investigation of the crystallinity of the substrate material could provide information about the contribution of crystallization to the non-linear behavior found in the influence of T_b. Future work should also focus on the combination of insert moulding and injection compression moulding to minimize the blister formation in FIM parts. Compression moulding has seen successfully used for the FIM process in several studies [29,30] and forms a strong basis for future research both for experimental and numerical activities.

Author Contributions: T.W.: Experimental analysis, Processing of the results and Writing. A.I.: Conceptualization, Writing, Reviewing and Editing. H.N.H.: Project administration, Reviewing and Editing. G.T.: Reviewing and Editing. B.R.W.: Reviewing and Editing. All authors have read and agreed to the published version of the manuscript.

Funding: This research received no external funding.

Conflicts of Interest: The authors declare no conflict of interest.

References

1. Kim, S.Y.; Kim, S.H.; Oh, H.J.; Lee, S.H.; Youn, J.R. Residual stress and viscoelastic deformation of film insert molded automotive parts. *J. Appl. Polym. Sci.* **2010**, *118*, 2530–2540. [CrossRef]
2. Kim, S.H.; Jeong, J.H.; Youn, J.R. Nanopattern insert molding. *Nanotechnology* **2010**, *21*, 205302. [CrossRef] [PubMed]
3. Weigelt, K.; Hambsch, M.; Karacs, G.; Zillger, T.; Hubler, A.C. Labeling the World: Tagging Mass Products with Printing Processes. *IEEE Pervasive Comput.* **2010**, *9*, 59–63. [CrossRef]
4. Leong, Y.W.; Umemura, T.; Hamada, H. Film insert molding as a novel weld-line inhibition and strengthening technique. *Polym. Eng. Sci.* **2008**, *48*, 2147–2158. [CrossRef]
5. Oh, H.J.; Song, Y.S.; Lee, S.H.; Youn, J.R. Development of Warpage and Residual Stresses in Film Insert Molded Parts. *Polym. Eng. Sci.* **2009**, *49*, 1389–1399. [CrossRef]
6. Kim, S.Y.; Sung, H.K.; Hwa, J.O.; Seung, H.L.; Soo, J.B.; Jae, R.Y.; Sung, H.L.; Sun, W.K. Molded geometry and viscoelastic behavior of film insert molded parts. *J. Appl. Polym. Sci.* **2009**, *111*, 642–650. [CrossRef]
7. Chen, H.L.; Chen, S.C.; Liao, W.H.; Chien, R.D.; Lin, Y.T. Effects of insert film on asymmetric mold temperature and associated part warpage during in-mold decoration injection molding of PP parts. *Int. Commun. Heat Mass Transf.* **2013**, *41*, 34–40. [CrossRef]
8. Kim, S.Y.; Lee, J.T.; Kim, J.Y.; Youn, J.R. Effects of film and substrate dimensions on warpage of film insert molded parts. *Polym. Eng. Sci.* **2010**, *50*, 1205–1213. [CrossRef]
9. Kim, J.Y.; Kim, S.Y.; Song, Y.S.; Youn, J.R. Relationship between the crystallization behavior and the warpage of film-insert-molded part. *J. Appl. Polym. Sci.* **2011**, *120*, 1539–1546. [CrossRef]
10. Wöhner, T.; Hansen, H.N.; Tosello, G.; Islam, A. Characterization method for blisters created in film insert moulding. In Proceedings of the Euspen's 15th International Conference & Exhibition, Leuven, Belgium, 1–6 June 2015.
11. Islam, A.; Hansen, H.; Bondo, M. Experimental investigation of the factors influencing the polymer-polymer bond strength during two-component injection moulding. *Int. J. Adv. Manuf. Technol.* **2010**, *50*, 101–111. [CrossRef]
12. Kim, S.Y.; Oh, H.J.; Kim, S.H.; Kim, C.H.; Lee, S.H.; Youn, J.R. Prediction of Residual Stress and Viscoelastic Deformation of Film Insert Molded Parts. *Polym. Eng. Sci.* **2008**, *48*, 1840–1847. [CrossRef]

13. Chen, S.C.; Li, H.M.; Huang, S.T.; Wang, Y.C. Effect of Decoration Film on Mold Surface Temperature during in-Mold Decoration Injection Molding Process. *Int. Commun. Heat Mass Transf.* **2010**, *37*, 501–505. [CrossRef]

14. Leong, Y.W.; Yamaguchi, S.; Mizoguchi, M.; Hamada, H.; Ishiaku, U.S.; Tsuji, T. The effect of molding conditions on mechanical and morphological properties at the interface of film insert injection molded polypropylene-film/polypropylene matrix. *Polym. Eng. Sci.* **2004**, *44*, 2327–2334. [CrossRef]

15. Leong, T.W.; Kotaki, M.; Hamada, H. Effects of the molecular orientation and crystallization on film–substrate interfacial adhesion in poly (ethylene terephthalate) film-insert moldings. *J. Appl. Polym. Sci.* **2007**, *104*, 2100–2107. [CrossRef]

16. Hales, J.H.; Kallesøe, C.; Lund-Olesen, T.; Johansson, A.; Fanøe, H.C.; Yu, Y.-j.; Lund, P.B.; Vig, A.L.; Tynelius, O.; Christensen, L.H. Micro fuel cells power the hearing aids of the future. *Fuel Cellls Bull.* **2012**, *2012*, 12–16. [CrossRef]

17. Wöhner, T.; Senkbeil, S.; Olesen, T.L.; Hansen, H.N.; Islam, A.; Tosello, G. Mould Design and Material Selection for Film Insert Moulding of Direct Methanol Fuel Cell Packaging. In Proceedings of the 4M/ICOMM2015 Conference, Milan, Italy, 31 March–2 April 2015; pp. 259–262.

18. Marhöfer, M.; Tosello, G.; Islam, A.; Hansen, H. Gate Design in Injection Molding of Microfluidic Components Using Process Simulations. *ASME J. Micro Nano-Manuf.* **2016**, *4*, 025001-1–025001-8.

19. Marhöfer, D.M.; Tosello, G.; Islam, A.; Hansen, H.N. Comparative analysis of different process simulation settings of a micro injection molded part featuring conformal cooling. In Proceedings of the Euspen's 15th International Conference and Exhibition, Leuven, Belgium, 1–5 June 2015; pp. 65–66.

20. Costa, F.S.; Tosello, G.; Whiteside, B.R. Best practice strategies for validation of micro moulding process simulation. In Proceedings of the Polymer Process Engineering Conference, Galati, Romania, 22–23 October 2009; pp. 259–261.

21. Doagou-Rad, S.; Islam, A.; Jensen, J.S.; Alnasser, A. Interaction of nanofillers in injectionmolded graphene/carbon nanotube reinforced PA66 hybrid nanocomposites. *J. Polym. Eng.* **2018**, *38*, 971–981. [CrossRef]

22. Moldflow Insight. In-Mold Labels Modeling (Procedure), Autodesk Knowledge Network. Available online: https://knowledge.autodesk.com/support/moldflow-insight/learn-explore/caas/CloudHelp/cloudhelp/2017/ENU/MoldflowInsight/files/GUID-A089CAEF-0F2B-4B90-8FF4-A2A79DF82080-htm.html (accessed on 10 February 2020).

23. Baek, S.J.; Kim, S.Y.; Lee, S.H.; Youn, J.R.; Lee, S.H. Effect of Processing Conditions on Warpage of Film Insert Molded Parts. *Fibers Polym.* **2008**, *9*, 747–754. [CrossRef]

24. Zheng, R.; Kennedy, P.K. A model for post-flow induced crystallization: General equations and predictions. *J. Rheol.* **2004**, *48*, 823. [CrossRef]

25. Moldflow Insight. Crystallization Analysis (Concept), Autodesk Knowledge Network. Available online: https://knowledge.autodesk.com/support/moldflow-insight/learn-explore/caas/CloudHelp/cloudhelp/2016/ENU/MFIA-WhatsNew/files/GUID-29607A41-680E-4A30-9F4B-8B86DEC8F117-htm.html (accessed on 8 April 2020).

26. Islam, A.; Li, X.; Wirska, M. Injection Moulding Simulation and Validation of Thin Wall Components for Precision Applications. In *Advances in Manufacturing II*; Springer: Berlin, Germany, 2019; pp. 96–107.

27. Lee, H.S.; Isayev, A.I. Numerical Simulation of Flow-Induced Birefringence: Comparison of Injection and Injection/Compression Molding. *Int. J. Precis. Eng. Manuf.* **2007**, *8*, 66–72.

28. Chen, S.C.; Chen, Y.C.; Peng, H.S.; Huang, L.T. Simulation of Injection Compression Molding Process. Part 3: Effect of process conditions on part birefringence. *Adv. Polym. Technol.* **2002**, *21*, 177–187. [CrossRef]

29. Lee, H.; Park, J. Experimental study of injection-compression molding of film insert molded plates. *Int. J. Precis. Eng. Manuf.* **2014**, *15*, 455–461. [CrossRef]

30. Lee, H.S.; Yoo, Y.G. Effects of Processing Conditions on Cavity Pressure during Injection-Compression Molding. *Int. J. Precis. Eng. Manuf.* **2012**, *13*, 2155–2161. [CrossRef]

micromachines

MDPI

Article

Hybrid Process Chain for the Integration of Direct Ink Writing and Polymer Injection Molding

Dario Loaldi [1,2] , **Leonardo Piccolo** [2] , **Eric Brown** [3] , **Guido Tosello** [1] , **Corey Shemelya** [3] **and Davide Masato** [2,*]

[1] Department of Mechanical Engineering, Technical University of Denmark, 2800 Kgs. Lyngby, Denmark; darloa@mek.dtu.dk (D.L.); guto@mek.dtu.dk (G.T.)

[2] Department of Plastics Engineering, University of Massachusetts Lowell, Lowell, MA 01854, USA; leonardo_piccolo@uml.edu

[3] Department of Electrical and Computer Engineering, University of Massachusetts Lowell, Lowell, MA 01854, USA; Eric_Brown2@student.uml.edu (E.B.); Corey_Shemelya@uml.edu (C.S.)

* Correspondence: Davide_Masato@uml.edu; Tel.: +1-978-934-2836

Received: 3 April 2020; Accepted: 14 May 2020; Published: 18 May 2020

Abstract: The integration of additive manufacturing direct-writing technologies with injection molding provides a novel method to combine functional features into plastic products, and could enable mass-manufacturing of custom-molded plastic parts. In this work, direct-write technology is used to deposit conductive ink traces on the surface of an injection mold. After curing on the mold surface, the printed trace is transferred into the plastic part by exploiting the high temperature and pressure of a thermoplastic polymer melt flow. The transfer of the traces is controlled by interlocking with the polymer system, which creates strong plastic/ink interfacial bonding. The hybrid process chain uses designed mold/ink surface interactions to manufacture stable ink/polymer interfaces. Here, the process chain is proposed and validated through systematic interfacial analysis including feature fidelity, mechanical properties, adhesion, mold topography, surface energy, and hot polymer contact angle.

Keywords: direct-writing; additive manufacturing; injection molding; micro manufacturing; functionalization

---

1. Introduction

Recent research efforts have opened new opportunities for the integration of Additive Manufacturing (AM) technologies with other mass-market fabrication techniques. The increasing industrial adoption of AM technologies has been shifting the focus from prototyping to direct integration with established processing technologies and applications. In particular, direct-write additive manufacturing has shown promise in a variety of applications ranging from radio frequency (RF) to sensing and even microfluidic integration [1–10]. Among AM technologies, direct-write micro-dispensing using silver conductive inks is one of the most common methods used to create printed electronics [11,12]. However, direct-write applications are no-longer limited to printed electronics, and inks are no-longer limited to silver-based nano particles. For example, Ruthenium and Barium Strontium Titanate inks have been developed for dielectric properties, Li4Ti5O12 (LTO) and LiFePO4 (LFP) have been used for printed batteries, BO_3Y (yttrium borate) was used as a printable luminescent material, and even bio-materials including silk fibroin has been printed with direct-write methods [13–26]. The material and process flexibility of direct-write AM offers significant opportunities for integration with mass-manufacturing technologies, such as plastic injection molding.

The current integration of functional features into plastic parts relies on the use of multi-step technologies, such as insert molding [27] or injection over-molding [28]. These technologies are characterized by high productivity, but limited flexibility for the integration of electronic features and functionalities. High production volumes are guaranteed through the separation of manufacturing in two consecutive steps: injection molding of the device and fabrication of the circuit traces.

Recent efforts to improve design flexibility have focused on post-processed methods. In particular, new process chains have been proposed to integrate printed circuits on injection molded parts using direct-write onto the plastic surface [29]. However, typical direct-write printing would require printing on difficult-to-reach areas, such as slots or internal surfaces to create complex geometries [30]. Moreover, the surface roughness, structures adhesion, and mechanical resistance of AM circuits can be critical factors in some applications, such as RF and sensing.

Here we propose the use of direct-writing AM technology to additively deposit conductive features on an injection mold surface. The printed traces undergo a transfer process into the polymeric part during the molding process. The volumetric integration of printed features into plastic parts is enabled by controlling the relative plastic/ink and mold/ink interfacial strengths. This process can also be further enhanced through the use of engineered injection-mold surface properties and coatings which have previously been demonstrated as an effective solution to control polymer flows both during the injection [31,32] and ejection phase [33].

To understand the printing of functional ink features onto injection mold materials, it is crucial to understand the interface interactions involved in the process. The creation of a quality ink/mold interface is the key parameter to achieve consistent printing. However, the ink/polymer interface should be stronger, to enable the final "transfer" to the plastic part. The ink/polymer interface is controlled by a variety of phenomena, including surface roughness [34], surface energy [35,36], and processing conditions [37]. In particular, high surface roughness has been shown to promote stronger interfacial interactions [38]. For these high roughness interfaces, the polymer melt has the capability of entering topographical voids, increasing the surface contact area between the melt and mold [39]. The investigation of these interfaces should also consider the differences in surface energy (i.e., polar and a-polar components) between two materials. The design of the material systems involved at the interfaces, allows achievement of stronger or weaker surface adhesion, depending on the similarities between polar and a-polar components of materials surface energy [40].

In this work, a hybrid molding process chain that enables the combination of ink dispensing technologies with conventional plastic injection molding is developed. Here we describe how nano-particle ink is printed, cured, and sintered before exposure to the polymer melt. After molding, the ink system of traces is embedded in the polymer parts and released from the mold surface due to a customized ink/polymer interface with differential polymer shrinkage. This work reports the characterization of the interactions at the mold/ink/polymer interfaces considering topography parameters, surface energy, and contact angle of the hot polymer melt. Additional mechanical testing is performed to evaluate the strength of molded devices and the adhesion of ink traces to the plastic parts produced with this method. It is expected that the advantages of integrating the embedded traces during molding will enable a stronger ink/polymer interface and improve surface homogeneity and finish compared to post-process printing methods.

2. Process Chain Development

Figure 1 describes the novel process chain and its main manufacturing steps. The process chain follows three primary processing steps: (i) mold preparation, (ii) direct-write AM, (iii) injection molding.

(i) Mold preparation: the mold was first pre-treated using a solvent/polymer solution, which was deposited on the surface where the ink will be printed. The releasing agent ensures high-quality printing and improves release to the polymer system during the injection molding process.

(ii) Direct-write AM: the printing process was carried out using an ink micro-dispensing system (Pro4, Nordson EDF, East Providence, RI, USA). The dispensing nozzle can be adjusted for multiple trace dimensions with the resulting feature sizes dependent on the selected ink, the deposition rate, and curing methods. Once printed, the ink was cured/sintered in a vacuum oven. It should be noted that other curing methods could be used including laser curing and photonic curing.

(iii) Injection molding: the mold, with the printed/sintered traces, was assembled into an injection molding machine for processing. The injected polymer melt flowed over the structures and the rapid cooling formed a strong bond between the melt and ink. The final polymer part with integrated traces was then ejected from the mold, resulting in a hybrid, composite part.

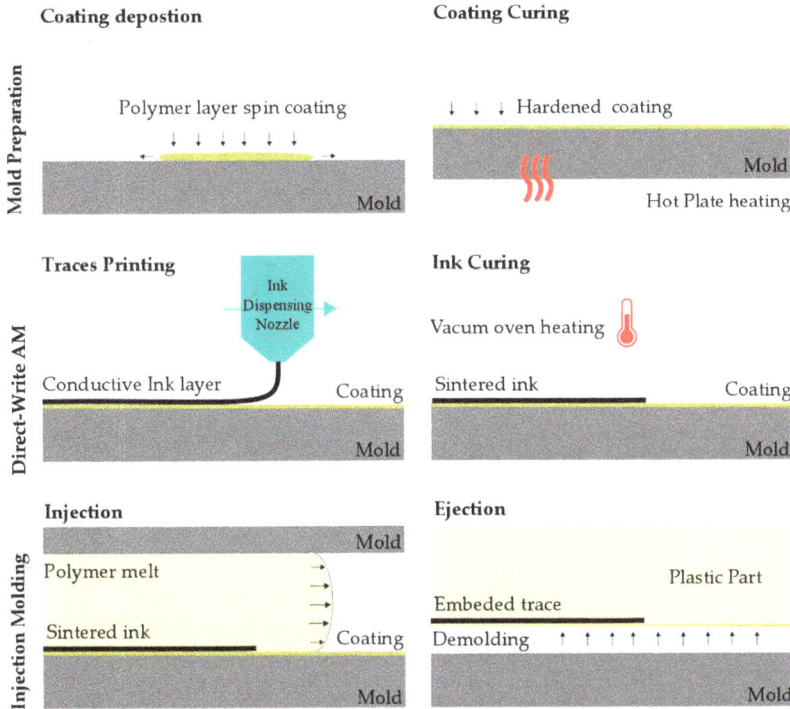

Figure 1. Steps of the process chain developed to manufacture molded interconnect devices.

3. Materials and Methods

3.1. Mold Surface Preparation

Before printing conductive ink onto the injection mold surface, the mold was pretreated with an ABS (Trilac® ABS-MP1000 Polymer Technology and Services, LLC (PTS), Heath, OH, USA) solvent solution. A typical coating process would include solution-based spin coating, alternatively, spray coating techniques to compensate for mold size and/or weight constraints.

The resulting mold surface maintains a thin, uniform, polymer coating directly on the stainless-steel. After coating, the mold is heated on a hot plate at 120 °C for 5 min to remove excess solvent from the coating. The surface treatment was designed and used to improve relative surface energies, enabling optimal mold/ink interface for molding "transfer."

3.2. Direct-Write Printing

A silver nanoparticle ink (CB 028, DuPont®, Wilmington, DE, USA) was selected for testing as it represents the standard ink-based material system in the field of printed electronics. As such, this investigation also represents a general proof-of-concept for eventual process integration for electrical applications.

Printing was carried out on the pre-treated mold surface using an automated micro-pen dispensing system (Nordson Pro4 EFD). Figure 2 shows a mold inside of the Nordson Pro4 EFD before printing. After printing, the mold with the printed traces was placed in a vacuum oven (Isotemp 282A, Thermo Fischer Scientific, Waltham, MA, USA) for sintering at 220 °C for 30 min.

To evaluate the effects of polymer pressure on ink/polymer interlocking and adhesion, printed traces were located at specific locations of the mold surface, based on the relative distance to the injection location. Each trace was printed in the form of a line with a design length of 20 mm, and a design height of 50 μm.

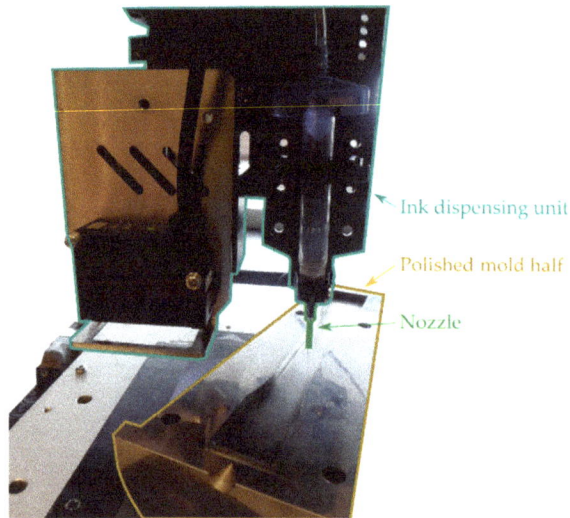

Figure 2. Ink dispensing system before printing over the injection mold surface.

3.3. Injection Molding Setup

Injection molding tests were run on a micro injection molding machine (Xplore® IM 12, Xplore Instruments BV, Sittard, The Netherlands). The machine features an injection cylinder of 10 mm in diameter with a maximum shot volume of 12 cm^3. The injection molding process conditions were set as follows: mold temperature at 80 °C, melt temperature at 260 °C, holding time of 8 s at 0.7 MPa. A single cavity stainless-steel mold was used to manufacture ASTM D638-14 type I tensile bars. During injection molding experiments, the mold with the printed traces was mounted on the machine and removed after every cycle for cleaning and additional printing.

The resin used for the experimentation was acrylonitrile butadiene styrene ABS (Trilac® ABS-MP1000 Polymer Technology and Services, LLC (PTS), Heath, OH, USA). This polymer was selected for its good processability, low viscosity, and rigidity. Additionally, this ABS resin is compatible with the coating layer that was used for mold pre-treatment before ink deposition.

3.4. Imaging and Topography Characterization

The dimensions of the traces printed on the mold surface and their qualitative inspection were performed using a stereomicroscope (SteREO Discovery V20, Zeiss, Oberkochen, Germany) to scan the whole printing area. Topographies were acquired using an optical profiler (Wyko NT2000, Bruker Nano, Tucson, AZ, USA), using a 20x magnification lens equipped with a Mireau interferometer. Figure 3a shows the characterization setup and the areas that were scanned.

From the acquired topographies, surface roughness *Sa* and mean summit curvature *Ssc* were evaluated according to ISO 25178:2012-2. These areal parameters were compared at multiple locations.

The surface amplitude roughness *Sa* was used as a quantitative indicator of amplitude while the average surface summit curvature *Ssc* characterizes the contact area between multiple surfaces.

The molded parts were sectioned to evaluate interface adhesion and morphology. Samples were embedded in a thermoset resin for polishing. For a smooth interface, the polishing process (Ecomet 250, Buehler, Lake Bluff, IL, USA) consisted of alternating pans of increasing mesh up to 3,000 with a total processing time of 20 min. Figure 3b shows a cross-section of a molded part with the integrated printed ink trace. The interface surfaces were qualitatively analyzed using Scanning Electron Microscopy (SEM - FEI, Quanta 400, Thermo Fisher Scientific, Waltham, MA, USA), allowing identification of voids, interface interlocking, and overall ink/polymer adhesion.

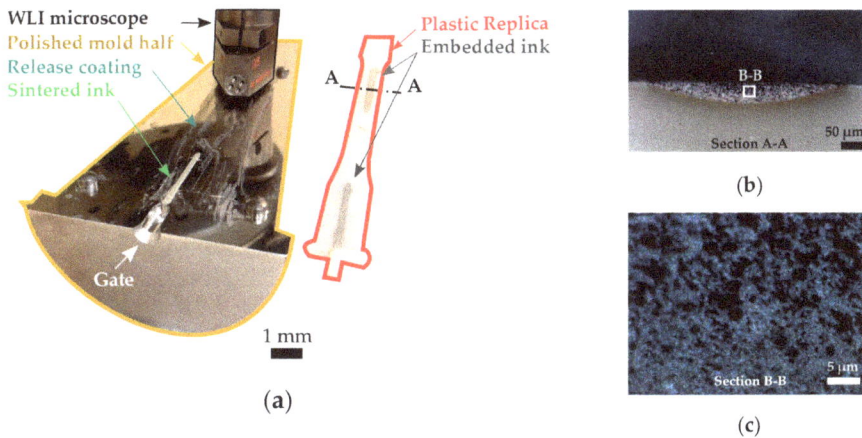

Figure 3. (a) Characterization of the mold before injection and the plastic replica after transfer. (b) Cross-section of the plastic part with the embedded traces, and (c) high magnification of the ink embedded in the printed traces.

3.5. Surface Energy and Polymer Melt Contact Angle

The adhesion at the ink/mold and ink/polymer interfaces was evaluated by quantifying the surface energy for the different materials involved in the process chain. Surface energy was measured using a drop shape analyzer (DSA 100, Krüss GmbH, Hamburg, Germany). Two different liquids (water and diiodomethane) were used for the estimation of surface energy based on the extended Fowkes' two liquids in contact model [41]. The measurements were performed on the following substrates: mold-polished steel, mold steel pre-treated for printing, sintered ink, and ABS molded parts in a trace-free area.

The effect of injection molding on the creation of a strong polymer/ink interface was evaluated by measuring the contact angle of the hot polymer melt over the mold surface. These tests were performed using the same drop shape analyzer on which a high temperature syringe dosing unit (TC21, Krüss GmbH, Hamburg, Germany) and measuring cell (TC3213, Krüss GmbH, Hamburg, Germany) were mounted. Contact angle measurements were performed using printed/sintered ink and hot ABS at injection molding melt temperature (i.e., 260 °C). The measurements were performed on polished steel.

3.6. Mechanical Properties

Mechanical properties of the injection molded parts with and without embedded traces were evaluated using standard ASTM dog-bone samples. The mechanical properties of the base ABS were benchmarked on ten samples manufactured following the procedure described in Section 3.3. Additionally, the mechanical properties were characterized for five samples manufactured with the proposed process chain.

Tensile testing (5966, Instron, Norwood, MA, USA) was carried out at room temperature according to ASTM D638. A load of 50 kN was used to achieve plastic fracture and the respective stress-strain curve. Young's modulus was calculated as the ratio of the stress at a percent strain of 0.2 %, while the Ultimate Tensile Strength (UTS) was evaluated as the maximum stress value before breakage.

3.7. Peeling Test

Further analysis of the molded parts with the integrated traces was carried out through a peeling test. The adhesion strength of the ink/plastic interface was evaluated according to ISO 2409:2013 cross-cut test. The test was performed using a hand-held, multi-blade cutting tool with a 1 mm cutting spacing between each blade. A pressure-sensitive adhesive tape (12.7 mm width, 600 HC-33 from 3M®) was used to remove the loose ink. The embedded ink trace were determined to have height of 200 μm, width of 2 mm, and length of 6 mm. The test consisted of sectioning the printed traces, attaching the tape to the surface of the hybrid part, and peeling. The result of the test is evaluated by inspecting the portions of the printed ink that are removed with the tape.

4. Results and Discussion

4.1. Influence of Surface Roughness

Figure 4 compares the measured surface features over the different surfaces described in the process chain. The results of the surface roughness evaluation showed significant differences between the mold, the release coating, and the sintered ink and are described in Table 1.

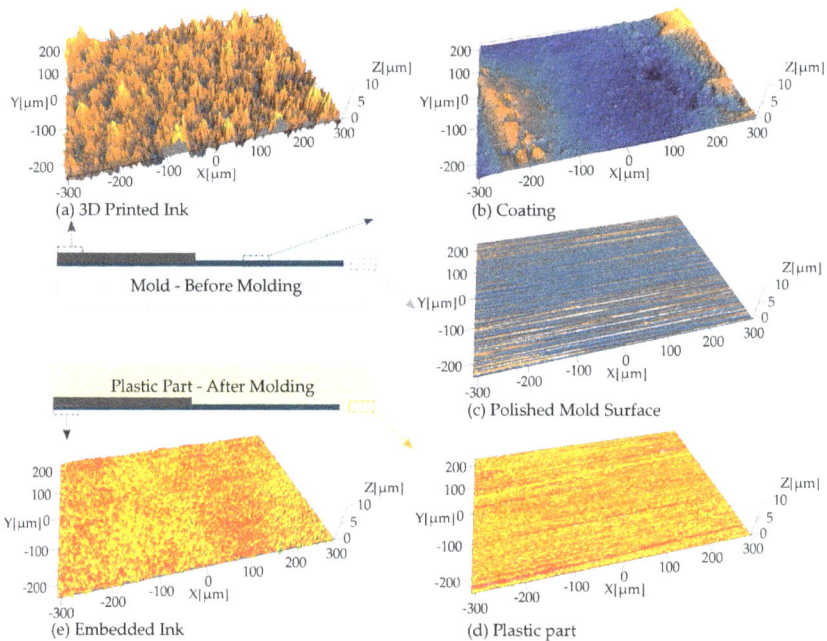

Figure 4. 3D view of the analyzed surface topographies: (**a**) 3D printed ink on the mold before injection molding, (**b**) coating on the mold before injection molding, (**c**) polished mold surface, (**d**) plastic part surface after demolding, (**e**) embedded ink on the plastic part.

Table 1. Summary table indicating *Sa* and *Ssc* parameter results for the analyzed topographies.

Surface	3D Printed Ink	Coating	Polished Mold Surface	Plastic Part	Embedded Ink
Sa/nm	1396	603	39	52	315
Ssc/µm	1.15	0.10	0.01	0.02	0.19

The printed ink surface had the highest roughness and summit curvature (cf. Figure 4a). Compared to the polished side of the mold, the ink shows a *Sa* higher by a factor 35x and an *Ssc* higher by a factor 115x. This is typical of direct-writing and AM technologies and is due to the particle-based nature of the inks. These traces are usually characterized by high surface roughness and their topography often exhibits inhomogeneity and macroscopic defects [42]. As shown in Figure 4e, embedding the traces during the molding process resulted in an average *Sa* value of 315 nm, which is 4x smaller than the printed ink trace alone (cf. Figure 4a). This roughness reduction confirms the effectiveness of the process chain in reducing the surface roughness of the printed electronics in the plastic part. Therefore, by printing the traces onto the mold and then transfer them via injection molding, it is possible to embed printed interconnects within a plastic part.

The SEM micrographs (Figure 5) of increasing magnification (500×–3000×–12000×) display the anisotropic nature and high roughness of the sintered ink surface before molding. As a comparison, Figure 6 shows the smoother topography resulting from the hybrid process chain presented in this work. Indeed, the traces printed on the injection mold surface are "flipped" when transferred to the plastic system in the injection molding process. The "transfer" process has the advantage of creating a smoother outer surface because of volumetric integration. This volumetric integration also is expected to result in higher wear resistance and higher resistance to mechanical shears.

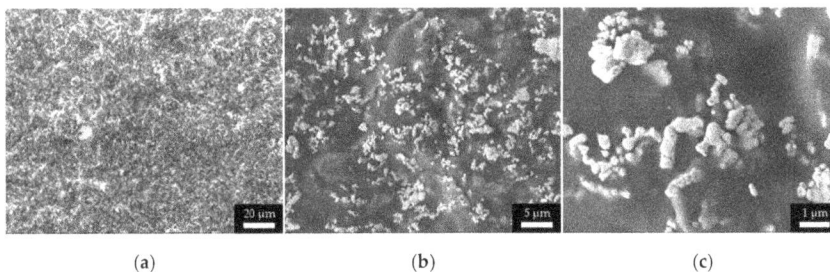

(a)	(b)	(c)

Figure 5. SEM micrographs at different magnification (500× (**a**) –3000× (**b**) –12000× (**c**)) of the sintered ink surface before injection molding. Micrographs were taken with an E beam power of 5 kV.

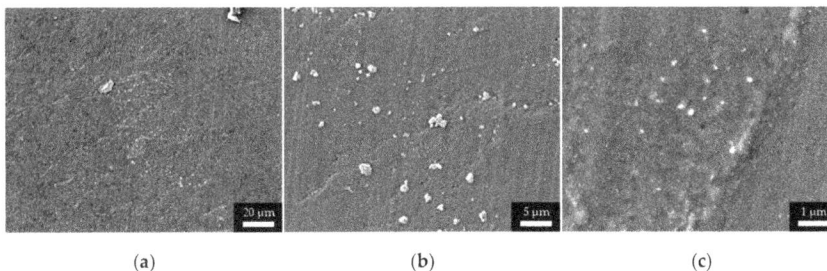

(a)	(b)	(c)

Figure 6. SEM micrographs at different magnification (500× (**a**) –3000× (**b**) –12000× (**c**)) of the sintered ink surface embedded in the injection molded part. Micrographs were taken with an E beam power of 5 kV.

Moreover, the high initial roughness of the printed traces, facing the polymer melt, favors interlocking upon injection molding. The molten polymer fills these topographical voids creating strong interactions at the polymer/ink interface. This exploits the high surface roughness typical of AM parts to generate higher interface adhesion and guarantees functionality.

4.2. Surface Energy

The surface energy of the different interfaces was used as an indirect measure of the adhesion strength between different materials interfaces [35]. By using contact angles measurements with a-polar liquid (water) and a-polar liquid (diiodiomethane) it was possible to calculate the total surface energy using the Fowkes' model [41]. From the contact angle measurements with the two liquids (Figure 7a), surface energy for different surface were obtained (Figure 7b). Then, the percentage of polar or disperse energy (Figure 7c) were analyzed to evaluate surface interactions.

Considering the experimental results, it was observed that surface energy increases significantly when reducing the surface roughness. In particular, the polished mold surface revealed a surface energy composed of 91% disperse and 9% polar contributions, which is double the polar affinity of an unpolished specimen. Similarly, the ink surface after curing is characterized by higher dispersion content, equal to 98% but now only a 2% polar contribution. Moreover, the overall surface energy in the polished mold surface is higher than the sintered ink, increasing from 32.8 ± 3.8 mJ/m^2 to 37.5 ± 6.5 mJ/m^2.

Contact angle measurements using the liquid ink (Figure 8a), suggest that the mold pre-treatment, used for release after injection molding (cf. Section 3.1), does not influence the direct-writing process, and thus the quality of ink deposition. However, the analysis of surface energy shows that higher surface affinity is formed between the mold and the plastic melt as opposed to the plastic and the sintered ink. From a process chain perspective, this measured surface affinity is very important to understand why the mold pre-treatment (cf. Section 3.2) favors the release and transfer mechanism [43]. In the absence of the pre-treatment layer, the ink would have a higher affinity with the mold steel, and transfer upon injection molding would not occur.

Contact angle measurements of the hot polymer melt were performed using a heated chamber. ABS samples were heated at 260 °C (i.e., molding temperature) and deposited on different surfaces to study the wetting behavior. The results indicate that a higher contact angle is formed over the sintered ink (i.e., 111.7°) and the unpolished mold surface (i.e., 116.4°) when compared to the polished mold surface (i.e., 103.2°). This contact angle adjustment demonstrates the effect of surface roughness on polymer/mold surface affinity, which is high for the sintered ink. As reported by Sorgato et al. the wettability of the polymer to a specific mold surface is crucial to determine the capability of the polymer to replicate the topography [36].

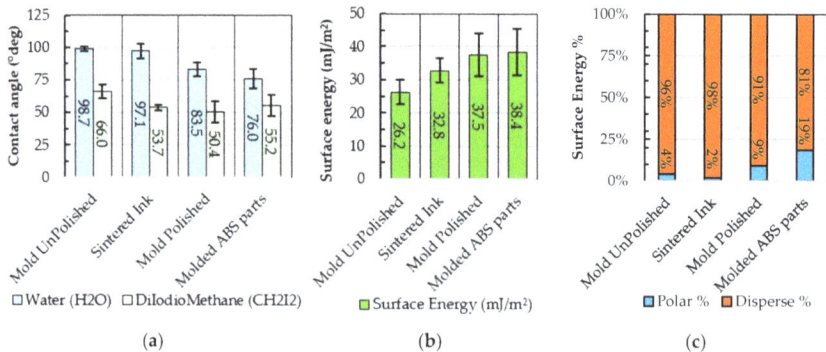

Figure 7. (a) Contact angle measurements for polar (H$_2$O) and a-polar (CH$_2$I$_2$) liquids: (b) surface energy for different surfaces; (c) percentage of dispersion and polar content for the analyzed surfaces.

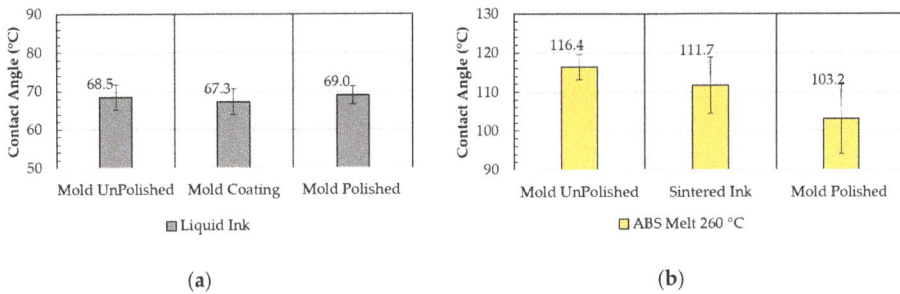

Figure 8. Contact angle measurements using (**a**) liquid ink and (**b**) ABS melt at 260 °C on different surfaces.

4.3. Mechanical Integrity and Interface Strength

The Young's modulus of elasticity, measured at 0.02% of elongation, and the Ultimate Tensile Strength (UTS) were used to analyze the effects of trace inclusion on the mechanical strength of the composite parts. The results are shown in Figure 9a,b, respectively. The average Young's modulus of the traditional parts is 3.1 ± 0.3 GPa, and the parts with embedded structures show a Young's modulus of 3.5 ± 0.2 GPa. This slight difference suggests that the embedded printed traces may act as a reinforcement, increasing the stiffness of the plastic part.

Analysis of the UTS values suggests that the printed traces embedded in the parts have no significant effect on the overall mechanical strength. The average for the ABS parts is 43.1 MPa, while the parts with the embedded traces have an average UTS of 43.2 MPa. This consistent UTS indicates that integration of the traces does not introduce any points of high-stress concentration, and the hybrid process chain has no negative effect on the tensile strength of the hybrid part.

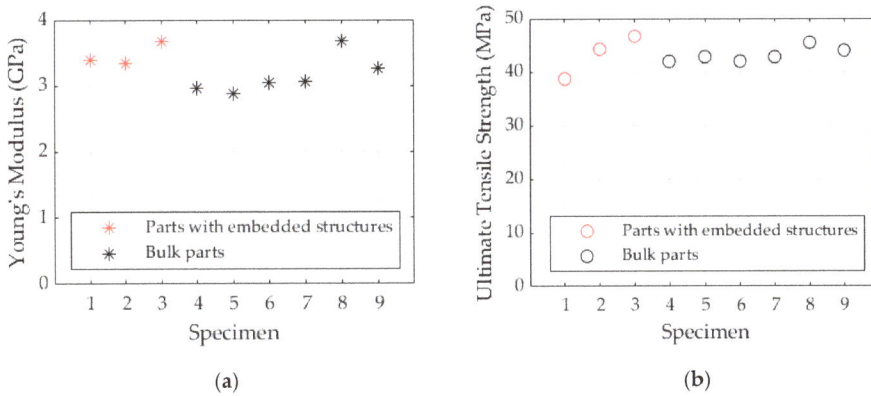

Figure 9. (**a**) Elastic modulus and (**b**) Ultimate Tensile Strength for samples with embedded printed electronics and bulk parts.

The adhesion at the ink/polymer interface for the parts was evaluated by carrying out peeling tests (Figure 10). The test was conducted on multiple traces located at multiple distances from the injection location, as described in Section 3.2. For all locations, strong adhesion was observed and no noticeable loss in adhesion could be measured. The results indicate that the proposed process chain can create consistent adhesion along the whole cavity where the polymer pressure is different.

Figure 10. Optical images of the embedded traces after the peeling test for a trace with a nominal width of (**a**) 250 μm and (**b**) 2000 μm.

5. Conclusions

This work presented an innovative process to integrate printed silver traces into injection molded plastic parts. This manufacturing process can be used to produce plastic parts that embed printed silver traces with custom surface roughness, unaltered mechanical properties, and strong polymer/electronic adhesion. The experimental results have demonstrated the effectiveness of the process in creating mechanical interlocking between the polymer system and the printed electronics.

The main findings of this work include:

- Topographical analysis of the different surfaces to verify the proposed process when compared to printing directly onto molded parts as a post-process. Specifically, the surface roughness of parts with integrated traces was reported to be low compared to that of post-processed, printed structures. The improved surface finish is explained by the intrinsic rotation of the structures during the "transfer" process.
- The high surface roughness of the printed structure before molding was exploited to create favorable interlocking at the ink/polymer interface, increasing adhesion. The trace transfer process from the mold/ink interface to the ink/polymer interface is enabled by exploiting the similar surface energy between the materials. Specifically, the results demonstrated that the surface energy affinity at the ink/polymer interface must be stronger than that of the ink/mold interface to promote transfer, indicating the need for a pre-treatment coating.
- It was expected that the composite molded parts would retain the mechanical stability, of bulk ABS parts. Our statistical analysis verified this effect, where volumetric integration of the printed traces into the plastic does not affect their structural properties. Moreover, peeling tests and micrographs of cross-sections demonstrate sufficient adhesion strength between the polymer and the ink.
- This work presents a first-in-class study on the feasibility, and methods required to integrate, additive direct-write tools with injection molding. The result is a hybrid technique enabling the fabrication of custom, design-on-demand plastic parts with integrated conductive structures. Future work will focus on the integration of additional additive techniques and the integration of more complex structures and molding techniques.

Author Contributions: Conceptualization, D.M. and C.S.; methodology, D.M., L.P., and D.L.; printing experiments D.L., E.B. and C.S.; injection molding experiments and characterization, D.L. and L.P.; analysis of experimental data, D.L., L.P., and D.M.; supervision, D.M. and G.T.; writing—original draft preparation, D.L.; writing—review and editing, D.M., G.T., and C.S. All authors have read and agreed to the published version of the manuscript.

Funding: This work was supported by the University of Massachusetts Lowell (Provost Office Start-Up funds to Prof. Masato and Prof. Shemelya).

Acknowledgments: The authors acknowledge the contribution of the Core Research Facilities at the University of Massachusetts Lowell for the help with parts characterization, and the Printed Electronics Research Collaborative for access to their printing facility. The authors acknowledge support from the Danish Innovation Fund (https://innovationsfonden.dk/en), in the research project of MADE DIGITAL, Manufacturing Academy of Denmark (http://en.made.dk/), Work Package WP3 "Digital manufacturing processes".

Conflicts of Interest: The authors declare no conflict of interest.

References

1. Shemelya, C.; Cedillos, F.; Aguilera, E.; Maestas, E.; Ramos, J.; Espalin, D.; Muse, D.; Wicker, R.; MacDonald, E. 3D printed capacitive sensors. In Proceedings of the 2013 IEEE Sensors, Baltimore, MD, USA, 3–6 November 2013; pp. 1–4.
2. Zadeh, E.; Sawan, M. High accuracy differential capacitive circuit for bioparticles sensing applications. In Proceedings of the 48th Midwest Symposium on Circuits and Systems, Covington, KY, USA, 7–10 August 2005; pp. 1362–1365. [CrossRef]
3. Shemelya, C.; Zemba, M.; Liang, M.; Espalin, D.; Kief, C.; Xin, H.; Wicker, R.; MacDonald, E. 3D printing multi-functionality: Embedded RF antennas and components. In Proceedings of the 2015 9th European Conference on Antennas and Propagation (EuCAP), Lisbon, Portugal, 13–17 April 2015; pp. 1–5.
4. Haghzadeh, M.; Armiento, C.; Akyurtlu, A. All-Printed Flexible Microwave Varactors and Phase Shifters Based on a Tunable BST/Polymer. *IEEE Trans. Microw. Theory Tech.* **2017**, *65*, 2030–2042. [CrossRef]
5. Ghafar-Zadeh, E.; Sawan, M.; Therriault, D.; Rajagopalan, S.; Chodavarapu, V.P. A direct-write microfluidic fabrication process for CMOS-based Lab-on-Chip applications. *Microelectron. Eng.* **2009**, *86*, 2104–2109. [CrossRef]
6. Khan, Y.; Thielens, A.; Muin, S.; Ting, J.; Baumbauer, C.; Arias, A.C. A New Frontier of Printed Electronics: Flexible Hybrid Electronics. *Adv. Mater.* **2020**, *32*, e1905279. [CrossRef] [PubMed]
7. Feng, J.Q.; Loveland, A.; Renn, M.J. Aerosol Jet Direct Writing Polymer-Thick-Film Resistors for Printed Electronics. *Preprints* **2020**, 2020020040. Available online: https://www.preprints.org/manuscript/202002.0040/v1 (accessed on 4 February 2020).
8. Huang, Q.; Zhu, Y. Printed Electronics: Printing Conductive Nanomaterials for Flexible and Stretchable Electronics: A Review of Materials, Processes, and Applications (Adv. Mater. Technol. 5/2019). *Adv. Mater. Technol.* **2019**, *4*, 1970029. [CrossRef]
9. Kwon, J.; Takeda, Y.; Shiwaku, R.; Tokito, S.; Cho, K.; Jung, S. Three-dimensional monolithic integration in flexible printed organic transistors. *Nat. Commun.* **2019**, *10*, 54. [CrossRef]
10. Zhang, Y.-Z.; Wang, Y.; Cheng, T.; Yao, L.-Q.; Li, X.; Lai, W.-Y.; Huang, W. Printed supercapacitors: Materials, printing and applications. *Chem. Soc. Rev.* **2019**, *48*, 3229–3264. [CrossRef]
11. Macdonald, E.; Salas, R.; Espalin, D.; Pérez, M.; Aguilera, E.; Muse, D.; Wicker, R. 3D Printing for the Rapid Prototyping of Structural Electronics. *IEEE Access* **2014**, *2*, 234–242. [CrossRef]
12. Seifert, T.; Sowade, E.; Roscher, F.; Wiemer, M.; Gessner, T.; Baumann, R.R. Additive Manufacturing Technologies Compared: Morphology of Deposits of Silver Ink Using Inkjet and Aerosol Jet Printing. *Ind. Eng. Chem. Res.* **2015**, *54*, 769–779. [CrossRef]
13. Haghzadeh, M.; Akyurtlu, A. All-printed, flexible, reconfigurable frequency selective surfaces. *J. Appl. Phys.* **2016**, *120*, 184901. [CrossRef]
14. Dardona, S.; Shen, A.; Tokgoz, C. Direct Write Fabrication of a Wear Sensor. *IEEE Sens. J.* **2018**, *18*, 3461–3466. [CrossRef]
15. Wang, J.; Pamidi, P.V.A.; Park, D.S. Screen-Printable Sol-Gel Enzyme-Containing Carbon Inks. *Anal. Chem.* **1996**, *68*, 2705–2708. [CrossRef] [PubMed]
16. Kosmala, A.; Wright, R.; Zhang, Q.; Kirby, P. Synthesis of silver nano particles and fabrication of aqueous Ag inks for inkjet printing. *Mater. Chem. Phys.* **2011**, *129*, 1075–1080. [CrossRef]
17. Kim, S.J.; Lee, J.; Choi, Y.-H.; Yeon, D.-H.; Byun, Y. Effect of copper concentration in printable copper inks on film fabrication. *Thin Solid Films* **2012**, *520*, 2731–2734. [CrossRef]
18. Bakhishev, T.; Subramanian, V. Investigation of Gold Nanoparticle Inks for Low-Temperature Lead-Free Packaging Technology. *J. Electron. Mater.* **2009**, *38*, 2720–2725. [CrossRef]

19. Gangwar, A.K.; Nagpal, K.; Kumar, P.; Singh, N.; Gupta, B.K. New insight into printable europium-doped yttrium borate luminescent pigment for security ink applications. *J. Appl. Phys.* **2019**, *125*, 074903. [CrossRef]

20. Tao, H.; Marelli, B.; Yang, M.; An, B.; Onses, M.S.; Rogers, J.A.; Kaplan, D.L.; Omenetto, F.G. Inkjet Printing of Regenerated Silk Fibroin: From Printable Forms to Printable Functions. *Adv. Mater.* **2015**, *27*, 4273–4279. [CrossRef]

21. Sun, K.; Wei, T.-S.; Ahn, B.Y.; Seo, J.Y.; Dillon, S.J.; Lewis, J.A. 3D Printing of Interdigitated Li-Ion Microbattery Architectures. *Adv. Mater.* **2013**, *25*, 4539–4543. [CrossRef]

22. Kamyshny, A.; Magdassi, S. Conductive nanomaterials for 2D and 3D printed flexible electronics. *Chem. Soc. Rev.* **2019**, *48*, 1712–1740. [CrossRef]

23. Dai, X.; Xu, W.; Zhang, T.; Shi, H.; Wang, T. Room temperature sintering of Cu-Ag core-shell nanoparticles conductive inks for printed electronics. *Chem. Eng. J.* **2019**, *364*, 310–319. [CrossRef]

24. Goh, G.L.; Saengchairat, N.; Agarwala, S.; Yeong, W.; Tran, T.A. Sessile droplets containing carbon nanotubes: A study of evaporation dynamics and CNT alignment for printed electronics. *Nanoscale* **2019**, *11*, 10603–10614. [CrossRef] [PubMed]

25. Kim, Y.Y.; Yang, T.; Suhonen, R.; Välimäki, M.; Maaninen, T.; Kemppainen, A.; Jeon, N.J.; Seo, J. Photovoltaic Devices: Gravure-Printed Flexible Perovskite Solar Cells: Toward Roll-to-Roll Manufacturing (Adv. Sci. 7/2019). *Adv. Sci.* **2019**, *6*, 1970044. [CrossRef]

26. Meng, L.; Zeng, T.; Jin, Y.; Xu, Q.; Wang, X. Surface-Modified Substrates for Quantum Dot Inks in Printed Electronics. *ACS Omega* **2019**, *4*, 4161–4168. [CrossRef] [PubMed]

27. Seong, G.S.; Lee, H.S. A study on the molding of dome shaped plastic parts embedded with electronic circuits. *J. Korea Soc. Die Mold Eng.* **2020**, *14*, 15–21.

28. Giusti, R.; Lucchetta, G. Analysis of the welding strength in hybrid polypropylene composites as a function of the forming and overmolding parameters. *Polym. Eng. Sci.* **2017**, *58*, 592–600. [CrossRef]

29. Lu, Y.; Yun, H.-Y.; Vatani, M.; Kim, H.; Choi, J.-W. Direct-print/cure as a molded interconnect device (MID) process for fabrication of automobile cruise controllers. *J. Mech. Sci. Technol.* **2015**, *29*, 5377–5385. [CrossRef]

30. Morais, M.V.C.; Reidel, R.; Weiss, P.; Baumann, S.; Hubner, C.; Henning, F. Integration of electronic components in the thermoplastic processing chain: Possibilities through additive manufacturing using conductive materials. In Proceedings of the 2018 13th International Congress Molded Interconnect Devices (MID), Würzburg, Germany, 25–26 September 2018; pp. 5–8. [CrossRef]

31. Masato, D.; Sorgato, M.; Babenko, M.; Whiteside, B.; Lucchetta, G.; Masato, D.; Marco, S.; Maksims, B.; Ben, W.; Giovanni, L. Thin-wall injection molding of polystyrene parts with coated and uncoated cavities. *Mater. Des.* **2018**, *141*, 286–295. [CrossRef]

32. Lucchetta, G.; Masato, D.; Sorgato, M.; Crema, L.; Savio, E. Effects of different mould coatings on polymer filling flow in thin-wall injection moulding. *CIRP Ann.* **2016**, *65*, 537–540. [CrossRef]

33. Sorgato, M.; Masato, D.; Lucchetta, G. Tribological effects of mold surface coatings during ejection in micro injection molding. *J. Manuf. Process.* **2018**, *36*, 51–59. [CrossRef]

34. Masato, D.; Sorgato, M.; Parenti, P.; Annoni, M.; Lucchetta, G. Impact of deep cores surface topography generated by micro milling on the demolding force in micro injection molding. *J. Mater. Process. Technol.* **2017**, *246*, 211–223. [CrossRef]

35. Packham, D. Surface energy, surface topography and adhesion. *Int. J. Adhes. Adhes.* **2003**, *23*, 437–448. [CrossRef]

36. Sorgato, M.; Masato, D.; Lucchetta, G. Effect of vacuum venting and mold wettability on the replication of micro-structured surfaces. *Microsyst. Technol.* **2016**, *23*, 2543–2552. [CrossRef]

37. Masato, D.; Sorgato, M.; Lucchetta, G. Analysis of the influence of part thickness on the replication of micro-structured surfaces by injection molding. *Mater. Des.* **2016**, *95*, 219–224. [CrossRef]

38. Lucchetta, G.; Borsato, F.; Bariani, P. Aluminum sheet surface roughness correlation with adhesion in polymer metal hybrid overmolding. *CIRP Ann.* **2011**, *60*, 559–562. [CrossRef]

39. Masato, D.; Sorgato, M.; Lucchetta, G. Effect of ultrasound vibration on the ejection friction in microinjection molding. *Int. J. Adv. Manuf. Technol.* **2018**, *96*, 1–14. [CrossRef]

40. Wu, S. *Polymer Interface and Adhesion*, 1st ed.; CRC Press: New York, NY, USA, 1982; p. 188.

41. Fowkes, F.M. attractive forces at interfaces. *Ind. Eng. Chem.* **1964**, *56*, 40–52. [CrossRef]

42. Beedasy, V.; Smith, P.J. Printed Electronics as Prepared by Inkjet Printing. *Materials* **2020**, *13*, 704. [CrossRef]
43. Lee, S.-H.; Shin, K.-Y.; Hwang, J.Y.; Kang, K.T.; Kang, H.S. Silver inkjet printing with control of surface energy and substrate temperature. *J. Micromech. Microeng.* **2008**, *18*, 75014. [CrossRef]

micromachines

MDPI

Article

Molecular Dynamics Simulations on the Demolding Process for Nanostructures with Different Aspect Ratios in Injection Molding

Can Weng, Dongjiao Yang and Mingyong Zhou *

College of Mechanical and Electrical Engineering, Central South University, Changsha 410083, China; canweng@csu.edu.cn (C.W.); 18198283005@163.com (D.Y.)
* Correspondence: zmy_csu@163.com; Tel.: +86-15974252435

Received: 26 August 2019; Accepted: 22 September 2019; Published: 23 September 2019

Abstract: Injection molding is one of the most potential techniques for fabricating polymeric products in large numbers. The filling process, but also the demolding process, influence the quality of injection-molded nanostructures. In this study, nano-cavities with different depth-to-width ratios (D/W) were built and molecular dynamics simulations on the demolding process were conducted. Conformation change and density distribution were analyzed. Interfacial adhesion was utilized to investigate the interaction mechanism between polypropylene (PP) and nickel mold insert. The results show that the separation would first happen at the shoulder of the nanostructures. Nanostructures and the whole PP layer are both stretched, resulting in a sharp decrease in average density after demolding. The largest increase in the radius of gyration and lowest velocity can be observed in 3:1 nanostructure during the separation. Deformation on nanostructure occurs, but nevertheless the whole structure is still in good shape. The adhesion energy gets higher with the increase of D/W. The demolding force increases quickly to the peak point and then gradually decreases to zero. The majority of the force comes from the adhesion and friction on the nanostructure due to the interfacial interaction.

Keywords: demolding process; molecular dynamics simulation; polypropylene; injection molding; nanostructure; depth-to-width ratio

1. Introduction

The functional surface with nanostructures exhibits excellent optical, electrochemical, and biological properties. Typical devices with the functional nanostructured surface, like super-hydrophobic coating, microfluidic chip, and antireflection film [1–4], are fabricated with the bottom-up methods, such as the self-assembly technology [5], or the top-down techniques, like electron beam lithography [1], ultra-precision milling [6], and nano-molding technology [7]. The fabrication quality of nanostructures plays an important role in its function implementation. Injection molding is one of the most potential techniques for fabricating polymer products in large numbers, with the advantages of cheaper and faster production. It is simple and diversified to fabricate nanostructures on the polymeric surface via injection molding because of the excellent workability, temperature resistance, and high modulus of elasticity [8].

Recently, surface structures with a higher aspect ratio or smaller dimension are being fabricated by injection molding [1,7,9,10]. During the injection molding process, the small dimension and high precision requirements would bring significant challenges, especially when the feature size is down to tens of nanometer or the aspect ratio is relatively high. The marked difference in thermal expansion coefficient between the metal insert and polymer material results in different shrinkages after demolding. The replication quality of nanostructures is quite sensitive to the change of processing parameters, which is not only determined by the filling process, but also the demolding process, due to the scale effect. Common demolding defects, including bending, necking, and structure fracture, would affect the mechanical properties. Moreover, non-destructive demolding is quite essential for achieving the performance of micro/nano-structured surfaces [11,12]. Therefore, it is necessary to analyze the interface behaviors at the nanoscale when the polymer is separated from the metallic mold insert.

The demolding force is mainly composed of adhesion, friction force, and shrinkage that are caused by different thermal expansion coefficients [12,13]. When the processing parameters were optimized to minimize the demolding force, it was found that the adhesion force had a higher influence than the friction force [14]. When considering the scale effect, the interaction mechanism between the polymer and mold insert at the nanoscale still requires further investigation. Experimental researches showed that, when the surface structure/roughness was down to several tens of nanometers, the strength of interfacial interaction, such as van der Waals force, was drastically increased, which forms a strong adhesion between polymer and metal insert [15]. Masato et al. [16] studied the influence of surface roughness of mold insert on the demolding force. If the surface roughness is less than 0.5 μm, the adhesion between the polymer and the insert dominates the final quality during separation.

Simulation methods that are based on continuum mechanics commonly fail to accurately predict the polymer behaviors at the nanoscale, because of the drastic changes in material properties. Recently, molecular dynamics (MD) simulation that is based on the atomistic movement theory provides the potential to study the interface behaviors at the atomic level [17,18]. The MD method has been successfully applied to the study the nano-imprinting [19–22] and injection molding process [23–25]. Even though some researchers are focusing on the demolding process in nano-imprinting, which could give inspiration and guidance to understand the demolding mechanism of polymeric nanostructures in nano-molding process, little literature on the demolding process of nanostructures in injection molding by the MD method is reported up to now. Despite the gaps in time-scale and size-scale between MD method and experimental research, the MD simulation can not only provide insight view of the movement behavior of polymer molecule, but also qualitatively estimate the morphology development in the nanoscale.

In this study, MD simulation models were constructed for the demolding investigation. Nano-cavities with different depth-to-width ratios (D/W) were built. Conformation changes, including molecular behaviors during the ejection, radius of gyration, and demolding velocity, were studied. Meanwhile, the density for both nanostructure and the whole polymer layer were compared to analyze the influence of the D/W. Adhesion energy and demolding force were proposed to understand the inner mechanism of the interfacial interaction between polymer and mold insert during the demolding process.

2. Materials and Methods

2.1. Materials and Model Constructing

The simulation system consisted of a polymer layer as the upper layer and a mold insert as the lower layer, as shown in Figure 1. For polymer layer, polypropylene (PP) was selected as the polymer material in nano-injection molding. An atomistic model of PP layer was constructed in a square box with dimensions of $6.0 \times 6.0 \times 5.3$ nm^3. The initial density of the layer was set to be 0.9 g/cm^3 at 293 K. There were a total of 200 chains in the box, and the degree of polymerization of each chain was 10. With such a low molecular weight, the nano-cavity can be better filled during the injection molding process, which would be helpful to build an initial model for the demolding simulation. Energy minimization and subsequently anneal treatment were utilized to optimize the conformation of the molecule structure in the PP layer. The layer was then heated up to 523 K to obtain a melt state, according to the actual condition in injection molding.

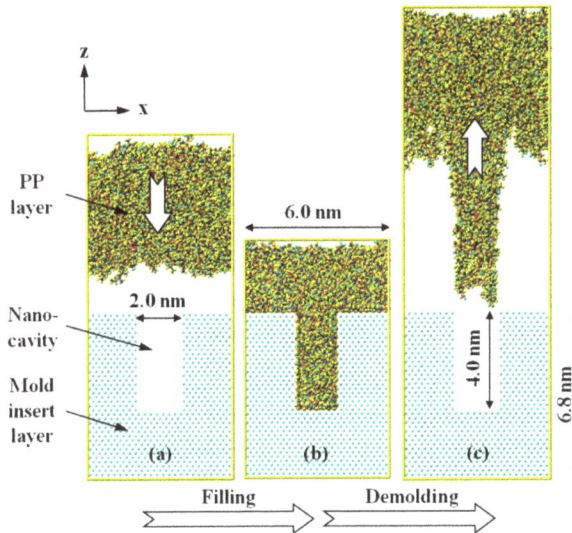

Figure 1. Atomistic model and procedure for the simulation. (**a**) The PP molecules fill the nano-cavity via injection molding, forming the nanostructure after cooling process in order to build (**b**) the initial model for demolding simulation. Afterwards, (**c**) the nanostructure is gradually pulled out from the mold insert.

The mold insert layer was composed of nickel atoms as a FCC structure with (1 0 0) plane. The insert layer had the same dimension in length and width as the PP layer. It is known that the mold insert has a much higher stiffness than the polymer. The insert layer was treated as a rigid body, so that all of the nickel atoms were constrained during the whole simulation. Rectangle nano-cavity was built in the upper of the nickel layer. The width of nano-cavity was set as a constant value of 2.0 nm, while the depths were 2.1 nm, 4.0 nm, and 6.1 nm, respectively. Three insert layers with different D/W were constructed, approximately from 1:1 to 3:1. Period boundary conditions in the x (length) and y (width) directions were used, while non-periodic and shrink-wrapped boundary condition in z (height) direction was set. Subsequently, the PP molecules can be shrinked or stretched during the simulation. The interaction between molecules in the upper region and the nickel atoms could be avoided by such a boundary condition.

2.2. Force Field and Simulation Procedure

Polymer Consistent Force Field (PCFF) was adopted to describe the intermolecular and non-bonded interactions between the atoms in the polymer layer. The force field PCFF was based on force field CFF91 with additional parameters being specified for polymer material. It consisted of not only the cross-term potentials, valence potentials, such as bond stretching, angular bending, and torsion potential, but also non-bonded interactions, including Lennard-Jones (9–6) and Coulomb potentials. The non-bonded interaction between PP molecules and nickel atoms was described with Lennard-Jones potential, with a cutoff distance of 1.25 nm.

Sufficient packing during the injection molding process is required in order to construct a demolding model with good replication quality in nano-cavity [26,27]. Hereby, a negative force (f_1) of 1.0 Kcal/mol·Å in the z-direction was applied to the PP layer. The insert layer was kept at a constant mold temperature of 373 K. As a result, the PP layer would gradually cool down to the mold temperature during the filling. The filling process was undertaken within a total of 4.0 ps in a constant particle number, volume, and temperature (NVT) ensemble, with a time step of 0.1 fs. The whole system was further cooled down from 373 K to 353 K in 1.0 ps. The final simulation result was treated as the initial model for the demolding process investigation. Afterwards, an external force (f_2) of 1.0 Kcal/mol·Å along z-axis was then applied to the PP layer to release the nanostructure from the insert layer in anotherr NVT ensemble. The demolding process was simulated in a varied time that depended on the D/W of the nano-cavity until the nanostructure was completely pulled out. All of the simulations mentioned above were performed by the Large-scale Atomic/Molecular Massively Parallel Simulator (LAMMPS) [28], which is an open source molecular dynamics package in a computer cluster.

3. Result and Discussion

3.1. Investigation on the Conformation Change

Figure 2 shows the snapshots of PP nanostructures with different D/W during the demolding process. The nano-cavity is fully filled during the injection molding process, as shown at 0 ps. It can be seen that there is hardly any separation in 1:1 nano-cavity at the early stage of the demolding process, while the upper molecules begin to show an upward movement. With the process goes on, molecules that are close to the nickel surface tend to fall behind because of the interaction with nickel atoms, as shown at 1.2 ps. Once the nanostructure was pulled out from the nano-cavity, boundary restriction from the insert layer no longer exists. Deformation in nanostructure gradually occurs and a slight expansion at the shoulder is eventually observed. This is mainly because of the interfacial interaction at the shoulder that was formed between PP and nickel atoms during the molding process. Some atoms in PP nanostructure are adhered to the cavity surface. Meanwhile, negative force is generated in molecules that are within the cutoff distance of non-bonded interaction, which thus results in surface unevenness and an elongation of the nanostructure. The nanostructure with D/W of 1:1 is completely released from the mold insert layer at a simulation time of 1.7 ps. With the increase of D/W, the demolding process becomes slower. Nanostructures are stretched during the pull-out process, regardless of the value of the D/W. It indicates that, in order to realize perfect replication fidelity of nanostructure, an anti-sticking treatment is suggested to be done before the demolding process in actual condition. For 2:1 and 3:1 nano-cavities, the separation first happens at the shoulder of the PP nanostructures, as shown at the simulation time of 0.6 ps in Figure 2b,c. Non-bonded interaction strength generated in the injection molding process is shown to be higher because there are more molecules being filled in the nano-cavity with the aspect ratio of 2:1 and 3:1. For 2:1 and 3:1 nano-cavities, the corresponding times for the complete separation are 2.7 ps and 3.6 ps, respectively. Although deformation in nanostructure can be observed during the ejection process, the integrity of each nanostructure is still in good shape generally.

The radius of gyration is commonly used as an indicator for the molecular size of polymer chains [29]. In this study, the mean square radius of gyration was introduced to analyze the conformation change of PP chains during the demolding process. It is demonstrated in Figure 3 that the radius first increases and then gradually seems to be equilibrated. It means that these chains are stretched away from the mold insert until the PP layer is completely released. Since more molecule chains are twisted in nano-cavity with 3:1 D/W, it implies greater flexibility for these chains. As a result, the largest change in the radius of gyration can be observed in this case. The radius of gyration reaches to 0.78 nm at the end of the demolding stage, being increased by 28.3%. Additionally, the increases in the radius of gyration are 23.1% for 2:1 nanostructure and 18.1% for 1:1 nanostructure, respectively. In addition, more simulation time is required to reach the equilibrium state when the depth of the nano-cavity is higher. Under combined effects of non-bonded interaction and the demolding force that was applied to the whole layer, the PP chains are greatly stretched, which makes good agreement with the conformation change in Figure 2.

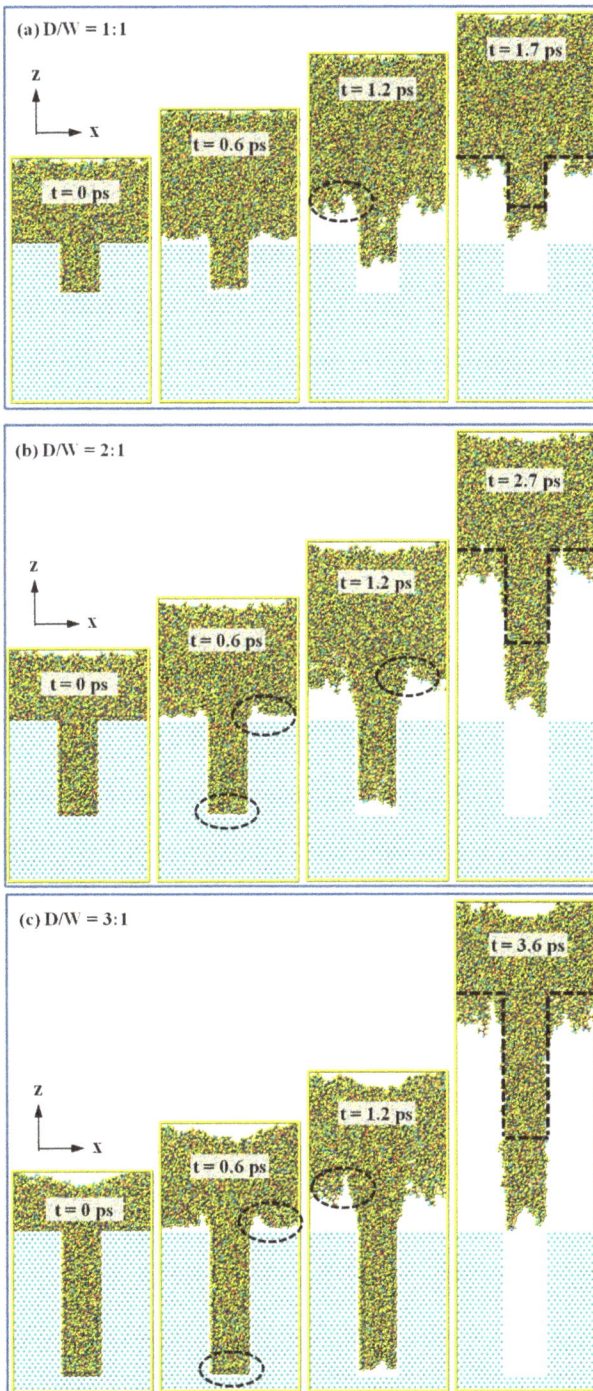

Figure 2. Snapshots of the PP nanostructure during the demolding process, with the D/W of these nano-cavities of (**a**) 1:1, (**b**) 2:1, and (**c**) 3:1, respectively.

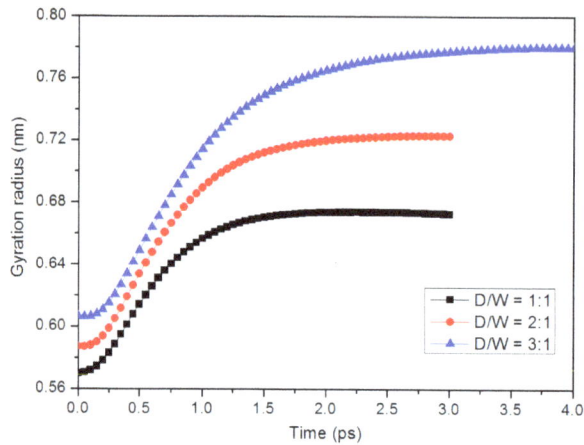

Figure 3. Gyration radii of polypropylene (PP) chain during the demolding process with different depth-to-width ratios (D/W).

Velocities in PP layer during the demolding process were calculated in order to investigate the elongation phenomenon of the injection-molded nanostructure. Figure 4 shows the profiles for the average velocities in both nanostructure and the whole layer. The velocity of the whole layer grows fast at the early stage and then reaches to a steady state eventually, approximately 2.9 nm/ps to 3.1 nm/ps. The velocity of PP nanostructure is lower than the whole layer since the PP molecules in nano-cavity are restrained by the interaction with nickel atoms. The velocity difference between nanostructure and the whole layer contributes to elongation growth in nanostructure's length. PP layer with higher D/W also shows a lower velocity during the demolding process because of the interfacial adhesion. By comparing the velocities of nanostructures with different D/W, it can be found that atoms in the 3:1 nanostructure move faster at the beginning of demolding. This is mainly because the adhesion energy that is generated by the non-bonded interaction is less than triple. Nevertheless, due to the spring back of PP molecules and the decrease of demolding force, the velocity grows fast with the separation of nanostructure in 1:1 nano-cavity. Section 3.3 will discuss further details regarding the interaction energy and demolding strength.

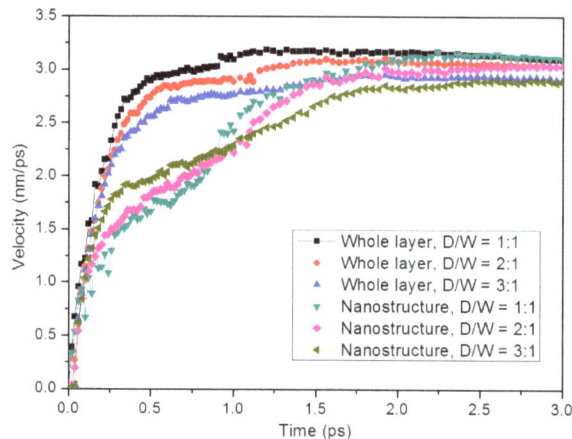

Figure 4. Velocities of the whole PP layer and the PP nanostructure during the demolding process with different D/W.

3.2. Determination of Density and Its Distribution

Density in nanostructure is calculated with the same height as the initial value before the demolding since the outline of PP nanostructure at the end of the demolding process is not so clear. It is demonstrated that both densities of nanostructures and the whole layer before the demolding were higher than the initial density (0.9 g/cm^3) due to the high pressure that was applied to the PP layer in injection molding. Polymer molecules are forced towards the insert layer, and the nano-cavity is highly filled. Similar results in the density change of the injection molded nanostructure were observed in our previous research [27]. The density of nanostructure remains almost the same, regardless of the aspect ratio. However, the average density dramatically drops when the nanostructure is completely released from the nano-cavity, less than half of the density before demolding, as shown in Figure 5. By analyzing the conformation changes in the radius of gyration and the snapshots during the demolding process, it can be found that the molecule chains in the whole PP layer were stretched. The total heights of both nanostructure and the PP layer are increased, while the corresponding mass almost remains the same.

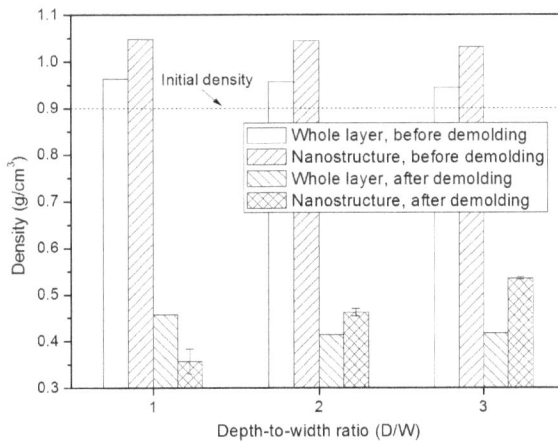

Figure 5. Densities of the whole PP layer and the PP nanostructure with different D/W, compared with the density values before and after demolding.

Figure 6 shows the density distribution of PP molecules in each slice along the *z*-axis, with a slice thickness of 0.2 nm and width of 6.0 nm. The density profile reaches a peak point at the top surface of the insert layer. It means that the PP molecules are enriched at the interface due to the packing pressure and the interfacial interaction between PP and nickel atoms. The interaction contributes to the adhesion force that prevents the PP molecules from separation. The density distribution at the interface further illustrates the reason for the deformation of the shoulder of PP nanostructure. The calculated slice density in PP nanostructure is much lower than the actual density because the width of the nanostructure is 2.0 nm while the width of the whole system is 6.0 nm, as shown in Figure 5. By comparing the density profiles before and after demolding, it can be observed that the total height of the nanostructure that is pulled out after demolding is obviously higher than the initial height, which means that the nanostructure is stretched after the demolding process. The density at the bottom of PP nanostructure shows the lowest value, while the density in the middle area of the nanostructure is relatively stable.

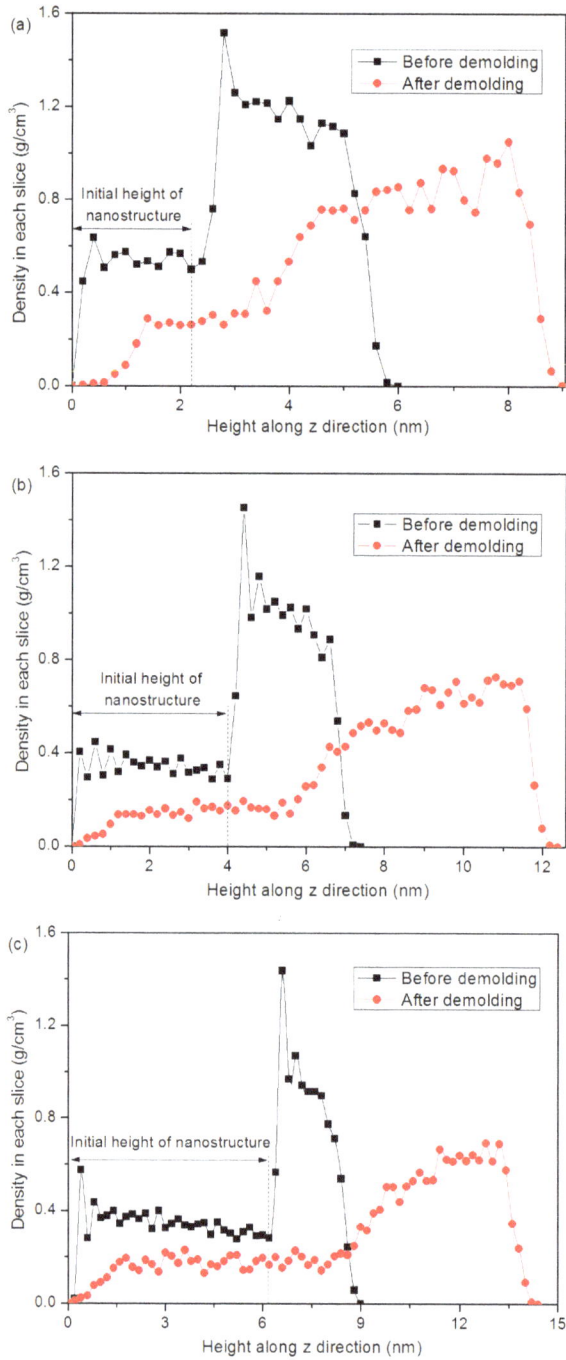

Figure 6. Density profiles in each slice of PP layer before and after the demolding process, with the D/W of these nano-cavities of (**a**) 1:1, (**b**) 2:1, and (**c**) 3:1, respectively.

3.3. Determination of Adhesion Energy and Demolding Force

According to the previous studies, the adhesion energy between the PP layer and nickel layer is generated by the non-bonded interaction at the interface [30–32]. In this study, the adhesion energy could be calculated from Equation (1)

$$E_{adhesion} = E_{interaction} = E_{total} - (E_{PP} + E_{Ni}) \tag{1}$$

where E_{total} is the total potential energy of the whole simulation system, E_{PP} is the potential energy of PP layer without any contribution from the mold insert, and E_{Ni} is the potential energy of the nickel layer without PP. The interaction energy is mainly determined by the close contacts between PP and nickel atoms that are within the cut-off distance (1.25 nm) of the non-bonded interaction. Figure 7 shows the adhesion energy distributions with different D/W during the demolding process. The negative value indicates that the PP and nickel atom attract to each other. The absolute value of the adhesion energy rapidly increases at the beginning, which is because the stable state of the initial conformation turns out to be destabilized due to the external releasing force f_2. The adhesion energy decreases shortly afterwards since the nanostructure is gradually pulled out from the nano-cavity. The effective contact area between PP and nickel atom is getting smaller during the demolding process. Maximum adhesion energy is found in the simulation system, with the aspect ratio of 3:1. The highest adhesion energy for 3:1 nano-cavity system is −3645.54 Kcal/mol, while the highest energies for 2:1 and 1:1 system are −3154.52 Kcal/mol and −2364.53 Kcal/mol, respectively. When the PP nanostructure is completely separated from the nano-cavity, the adhesion energy turns to zero, which means that the non-bonded interaction no longer exists.

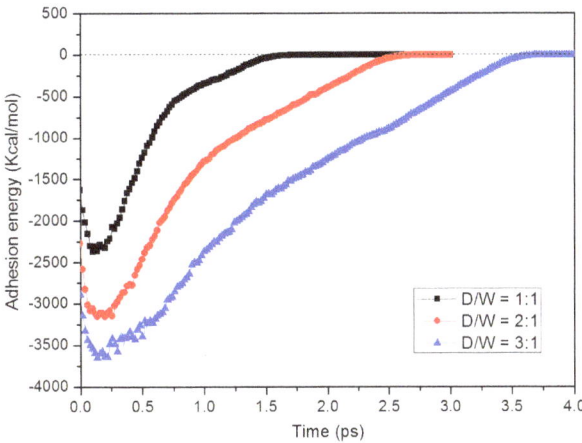

Figure 7. Adhesion energies during the demolding process with different D/W.

The total demolding force was investigated by calculating the inner force for each atom. As shown in Figure 8, the demolding force quickly increases to its peak point, prior to the separation of the nanostructure. Afterwards, it gradually decreases to zero shortly when the nanostructure is completely pulled out. Maximum demolding force is observed between the simulation time of 0.3 ps and 0.4 ps. During this period, most of the PP nanostructure is still in the nano-cavity, generating strong adhesion to the surface of both nano-cavity and mold insert. With the increase of D/W, a higher demolding force is generated, since there are more PP molecules surrounding around the interface. Figure 9 shows an example of the density contour for the whole atoms, with the D/W of 3:1 during the demolding process. At 0.4 ps, atoms in the nanostructure show relatively higher demolding force, especially these atoms in the upper area of the nanostructure, due to the strong adhesion. The demolding force per atom decreases at 1.2 ps, while the certain area in nanostructure still shows high demolding force, as shown in Figure 9b. When the nanostructure is pulled out, the demolding force falls to zero correspondingly.

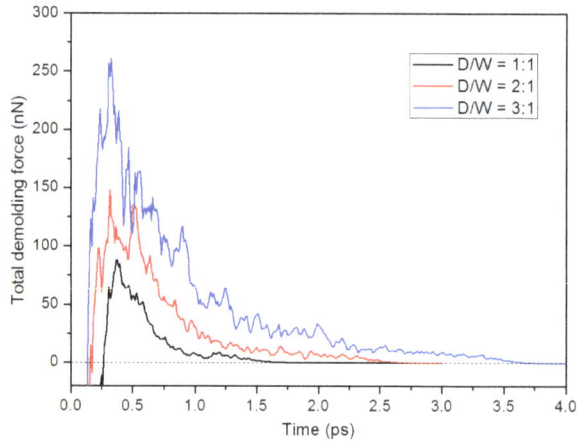

Figure 8. Total demolding forces of PP layer during the separation, with the D/W values of (lower curve) 1:1, (middle curve) 2:1, and (upper curve) 3:1.

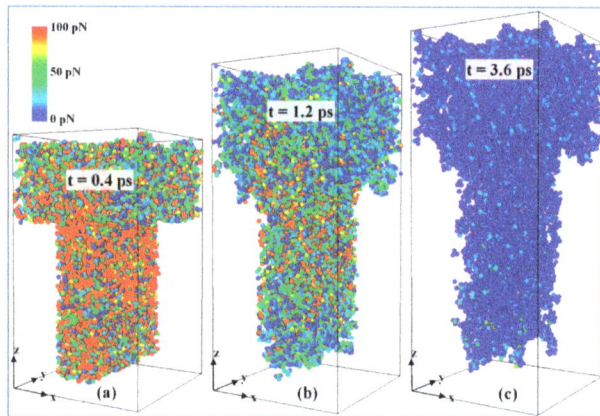

Figure 9. Demolding force per atom in 3:1 nanostructure at a simulation time of (**a**) 0.4 ps, (**b**) 1.2 ps, and (**c**) 3.6 ps.

The total demolding force and the average force were compared during the simulation between 0.3 ps and 0.4 ps in order to better understand the force distribution at the peak point. These forces on each atom of the whole PP layer and the nanostructure were calculated. It is demonstrated in Figure 10a that the total demolding force for the whole PP layer are 71.56 nN, 116.12 nN, and 222.23 nN for 1:1, 2:1, and 3:1 nano-cavity systems, while the total demolding force for PP nanostructure are 47.76 nN, 114.11 nN, and 213.06 nN, respectively. It is demonstrated that the majority of the demolding force comes from the adhesion and friction on the nanostructure, especially when the values of the D/W are 2:1 and 3:1. Although the average force per atom on the nanostructure has a relatively high standard deviation, the demolding force is nevertheless much higher than the average value of the whole layer. With the increases of D/W, total demolding force and average force on each atom both increase. It means that a higher resistance is generated at that time. Similar results on the demolding force distribution can also be observed in other simulation times.

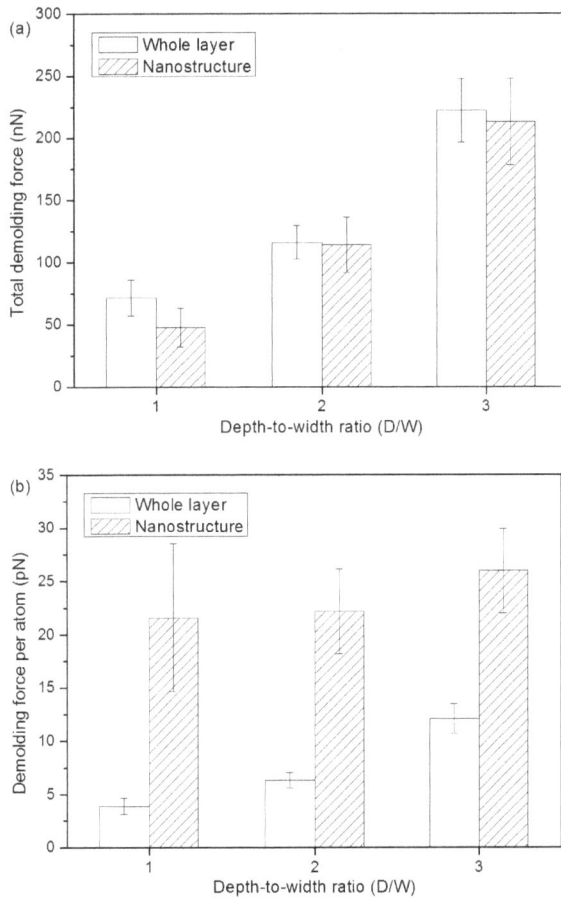

Figure 10. Demolding forces for different D/W, (**a**) the total demolding force and (**b**) the force per atom were calculated as an average value during the simulation time from 0.3 to 0.4 ps, when the highest demolding force occurs.

4. Conclusions

In this present work, molecular dynamics simulation system that consisted of a PP layer and a nickel mold insert layer was constructed to study the demolding process of the nanostructure in injection molding. The simulation results show that there is hardly any separation of PP molecules in the nano-cavity at the early demolding stage. The separation first happens at the shoulder of the nanostructure, especially for the structure with a high D/W. The nanostructure is stretched since the velocity of PP nanostructure is relatively lower than the whole PP layer. Analyzing the radius of gyration also stretches the PP chains in the whole layer. Moreover, an obvious drop in density is found after the demolding. During the separation, the largest increase in the radius of gyration is approximately 28.3%. The lowest velocity is observed in the 3:1 nanostructure due to the strong interfacial adhesion. Nanostructure deformation gradually occurs, which results in a slight expansion at the shoulder eventually. When the D/W is higher, the adhesion energy that is generated by the non-bonded interaction is higher. It means more demolding time is required for the complete separation. The demolding force increases quickly to its peak point before the separation of the nanostructure. Maximum force can be observed between 0.3 ps and 0.4 ps, with the total force of 71.56 nN, 116.12 nN, and 222.23 nN for 1:1, 2:1, and 3:1 nano-cavity systems. Atoms in nanostructure, especially in the upper area, show relatively higher demolding force. The majority of the force comes from the adhesion and friction on the nanostructure due to the interfacial interaction. With the increases of D/W, total demolding force and demolding force per atom are both higher.

Author Contributions: Proposed the idea of simulation process, C.W.; performed the MD simulations, D.Y.; analyzed the data and wrote the paper, M.Z.; offered the tutorials and suggestions, C.W.

Acknowledgments: This research was supported by the National Natural Science Foundation of China (Grant No: 51775562, 51805550).

Conflicts of Interest: The authors declare no conflict of interest.

References

1. Stormonth-Darling, J.M.; Pedersen, R.H.; How, C.; Gadegaard, N. Injection moulding of ultra high aspect ratio nanostructures using coated polymer tooling. *J. Micromech. Microeng.* **2014**, *24*, 075019. [CrossRef]
2. Utko, P.; Persson, F.; Kristensen, A.; Larsen, N.B. Injection molded nanofluidic chips: Fabrication method and functional tests using single-molecule DNA experiments. *Lab Chip* **2011**, *11*, 303–308. [CrossRef] [PubMed]
3. Zhou, M.; Xiong, X.; Jiang, B.; Weng, C. Fabrication of high aspect ratio nanopillars and micro/nano combined structures with hydrophobic surface characteristics by injection molding. *Appl. Surf. Sci.* **2018**, *427*, 854–860. [CrossRef]
4. Yanagishita, T.; Kondo, T.; Masuda, H. Preparation of renewable antireflection moth-eye surfaces by nanoimprinting using anodic porous alumina molds. *J. Vac. Sci. Technol. B* **2018**, *36*, 031802. [CrossRef]
5. Park, J.M. Fabrication of various nano-structured nickel stamps using anodic aluminum oxide. *Microsyst. Technol.* **2014**, *20*, 2157–2163. [CrossRef]
6. Weng, C.; Lee, W.B.; To, S. A study of the relevant effects on the maximum residual stress in the precision injection moulding of microlens arrays. *J. Micromech. Microeng.* **2010**, *20*, 035033. [CrossRef]
7. Kimura, F.; Kadoya, S.; Kajihara, Y. Effects of molding conditions on injection molded direct joining using a metal with nano-structured surface. *Precis. Eng.* **2016**, *45*, 203–208. [CrossRef]
8. Saarikoski, I.; Suvanto, M.; Pakkanen, T.A. Modification of polycarbonate surface properties by nano-, micro-, and hierarchical micro–nanostructuring. *Appl. Surf. Sci.* **2009**, *255*, 9000–9005. [CrossRef]
9. Matschuk, M.; Larsen, N.B. Injection molding of high aspect ratio sub-100 nm nanostructures. *J. Micromech. Microeng.* **2013**, *23*, 025003. [CrossRef]
10. Xie, H.; Huang, H.X.; Peng, Y.J. Rapid fabrication of bio-inspired nanostructure with hydrophobicity and antireflectivity on polystyrene surface replicating from cicada wings. *Nanoscale* **2017**, *9*, 11951–11958. [CrossRef]
11. Guo, Y.; Liu, G.; Xiong, Y.; Tian, Y. Study of the demolding process—Implications for thermal stress, adhesion and friction control. *J. Micromech. Microeng.* **2007**, *17*, 9–19. [CrossRef]
12. Guo, Y.; Liu, G.; Zhu, X.; Tian, Y. Analysis of the demolding forces during hot embossing. *Microsyst. Technol.* **2006**, *13*, 411–415. [CrossRef]
13. Dirckx, M.E.; Hardt, D.E. Analysis and characterization of demolding of hot embossed polymer microstructures. *J. Micromech. Microeng.* **2011**, *21*, 085024. [CrossRef]
14. Su, Q.; Gilchrist, M.D. Demolding forces for micron-sized features during micro-injection molding. *Polym. Eng. Sci.* **2016**, *56*, 810–816. [CrossRef]
15. Sasaki, T.; Koga, N.; Shirai, K.; Kobayashi, Y.; Toyoshima, A. An experimental study on ejection forces of injection molding. *Precis. Eng.* **2000**, *24*, 270–273. [CrossRef]
16. Masato, D.; Sorgato, M.; Parenti, P.; Annoni, M.; Lucchetta, G. Impact of deep cores surface topography generated by micro milling on the demolding force in micro injection molding. *J. Mater. Process. Technol.* **2017**, *246*, 211–223. [CrossRef]
17. Song, Q.; Ji, Y.; Li, S.; Wang, X.; He, L. Adsorption behavior of polymer chain with different topology structure at the polymer-nanoparticle interface. *Polymers* **2018**, *10*, 590. [CrossRef] [PubMed]
18. Hirai, Y.; Konishi, T.; Yoshikawa, T.; Yoshida, S. Simulation and experimental study of polymer deformation in nanoimprint lithography. *J. Vac. Sci. Technol. B* **2004**, *22*, 3288. [CrossRef]

19. Yang, S.; Yu, S.; Cho, M. Influence of mold and substrate material combinations on nanoimprint lithography process: Md simulation approach. *Appl. Surf. Sci.* **2014**, *301*, 189–198. [CrossRef]

20. Takai, R.; Yasuda, M.; Tochino, T.; Kawata, H.; Hirai, Y. Computational study of the demolding process in nanoimprint lithography. *J. Vac. Sci. Technol. B* **2014**, *32*, 06FG02. [CrossRef]

21. Kang, J.-H.; Kim, K.-S.; Kim, K.-W. Molecular dynamics study of pattern transfer in nanoimprint lithography. *Tribol. Lett.* **2007**, *25*, 93–102. [CrossRef]

22. Carrillo, J.M.; Dobrynin, A.V. Molecular dynamics simulations of nanoimprinting lithography. *Langmuir* **2009**, *25*, 13244–13249. [CrossRef] [PubMed]

23. Pina-Estany, J.; García-Granada, A.A. Molecular dynamics simulation method applied to nanocavities replication via injection moulding. *Int. Commun. Heat Mass Transf.* **2017**, *87*, 1–5. [CrossRef]

24. Zhou, M.; Jiang, B.; Weng, C. Molecular dynamics study on polymer filling into nano-cavity by injection molding. *Comput. Mater. Sci.* **2016**, *120*, 36–42. [CrossRef]

25. Dai, C.-F.; Chang, R.-Y. Molecular dynamics simulation of thread break-up and formation of droplets in nanoejection system. *Mol. Simulat.* **2009**, *35*, 334–341. [CrossRef]

26. Lin, H.-Y.; Chang, C.-H.; Young, W.-B. Experimental and analytical study on filling of nano structures in micro injection molding. *Int. Commun. Heat Mass Transf.* **2010**, *37*, 1477–1486. [CrossRef]

27. Zhou, M.; Xiong, X.; Drummer, D.; Jiang, B. Molecular dynamics simulation and experimental investigation of the geometrical morphology development of injection-molded nanopillars on polymethylmethacrylate surface. *Comput. Mater. Sci.* **2018**, *149*, 208–216. [CrossRef]

28. Plimpton, S. Fast parallel algorithms for short-range molecular dynamics. *J. Comput. Phys.* **1995**, *117*, 1–19. [CrossRef]

29. Kim, S.; Lee, D.E.; Lee, W.I. Molecular dynamic simulation on the effect of polymer molecular size in thermal nanoimprint lithographic (t-nil) process. *Tribol. Lett.* **2013**, *49*, 421–430. [CrossRef]

30. Prathab, B.; Subramanian, V.; Aminabhavi, T.M. Molecular dynamics simulations to investigate polymer–polymer and polymer–metal oxide interactions. *Polymer* **2007**, *48*, 409–416. [CrossRef]

31. Kisin, S.; Božović Vukić, J.; van der Varst, P.G.; de With, G.; Koning, C.E. Estimating the polymer–metal work of adhesion from molecular dynamics simulations. *Chem. Mater.* **2007**, *19*, 903–907. [CrossRef]

32. Liu, F.; Hu, N.; Ning, H.; Liu, Y.; Li, Y.; Wu, L. Molecular dynamics simulation on interfacial mechanical properties of polymer nanocomposites with wrinkled graphene. *Comput. Mater. Sci.* **2015**, *108*, 160–167. [CrossRef]

micromachines

MDPI

Article

Experimental Validation of Injection Molding Simulations of 3D Microparts and Microstructured Components Using Virtual Design of Experiments and Multi-Scale Modeling

Dario Loaldi⬮, Francesco Regi, Federico Baruffi, Matteo Calaon⬮, Danilo Quagliotti⬮, Yang Zhang and Guido Tosello *⬮

Department of Mechanical Engineering, Technical University of Denmark, Building 427A, Produktionstorvet, DK-2800 Kgs Lyngby, Denmark; darloa@mek.dtu.dk (D.L.); fr@rel8.dk (F.R.); federico.baruffi.91@gmail.com (F.B.); mcal@mek.dtu.dk (M.C.); danqua@mek.dtu.dk (D.Q.); yazh@mek.dtu.dk (Y.Z.)
* Correspondence: guto@mek.dtu.dk

Received: 3 June 2020; Accepted: 23 June 2020; Published: 24 June 2020

Abstract: The increasing demand for micro-injection molding process technology and the corresponding micro-molded products have materialized in the need for models and simulation capabilities for the establishment of a digital twin of the manufacturing process. The opportunities enabled by the correct process simulation include the possibility of forecasting the part quality and finding optimal process conditions for a given product. The present work displays further use of micro-injection molding process simulation for the prediction of feature dimensions and its optimization and microfeature replication behavior due to geometrical boundary effects. The current work focused on the micro-injection molding of three-dimensional microparts and of single components featuring microstructures. First, two virtual a studies were performed to predict the outer diameter of a micro-ring within an accuracy of 10 μm and the flash formation on a micro-component with mass a 0.1 mg. In the second part of the study, the influence of microstructure orientation on the filling time of a microcavity design section was investigated for a component featuring micro grooves with a 15 μm nominal height. Multiscale meshing was employed to model the replication of microfeatures in a range of 17–346 μm in a Fresnel lens product, allowing the prediction of the replication behavior of a microfeature at 91% accuracy. The simulations were performed using 3D modeling and generalized Navier–Stokes equations using a single multi-scale simulation approach. The current work shows the current potential and limitations in the use of micro-injection molding process simulations for the optimization of micro 3D-part and microstructured components.

Keywords: modeling; micro-injection molding; micro replication; process simulation

1. Introduction

A consolidated trend in micro-manufacturing consists of the adoption of replication technologies for large-scale productions. Due to its high throughput and overall capabilities, combined with the possibility of automating the process, micro-injection molding (μIM) is the most commonly found replication process in multiple applications and industries including medical, optical, consumer, sensors, and micro electro-mechanical systems (MEMS) [1]. To hasten the micro product design phase, optimize μIM process conditions as well as predict process quality and performance, significant attention has been dedicated to the numerical modeling and the simulation of such technology, aiming for the establishment of a μIM digital twin. The μIM process encompasses three families of products that significantly constrain the technical equipment required for processing and simulation [2], as described in Table 1. µIM process simulation has been developed from conventional IM modeling methods. Multiple commercially available software for the scope includes, for example, Autodesk Moldflow® (Autodesk Inc., San Rafael, CA, USA), Moldex3D (CoreTech System Co., Hsinchu County, Taiwan), Simpoe-Mold (Dassault Systèms, Hertogenbosch, Netherlands), Sigmasoft® (SIGMA Engineering GmbH, Aachen, Germany), and Rem3D® (Transvalor S.A., Mougins, France). The numerical simulation is structured by solving a system of equations that enclose the conservation of mass (Equation (1)), linear momentum (Equation (2)), and energy (Equation (3)).

$$\frac{d}{dt}\rho + \nabla \cdot (\rho v) = 0 \tag{1}$$

$$\rho \frac{d}{dt}v = \rho g - \nabla P + \eta \nabla^2 v \tag{2}$$

$$\rho c_p \left(\frac{\partial}{\partial t}T + v\nabla T \right) = \beta T \left(\frac{\partial}{\partial t}P + \vec{v}' \times \vec{\nabla} P \right) + \nabla \cdot (k\nabla T) + \eta \dot{\gamma}^2 \tag{3}$$

where ρ is the density; t is the time; v is the velocity vector; g the gravitational acceleration constant; P is the hydrostatic pressure; η is the viscosity; c_p is the specific heat; T is the temperature; β is the heat expansion coefficient; k is the thermal conductivity; and γ is the shear rate. Additional boundary conditions that describe the polymer flow are added to the system, which include the polymer pvT constitutive relationship often described with the Tait equation, and a velocity-dependent viscosity model often described with the Cross-WLF [3]. When scaling down from the macro IM toward µIM, several additional physics should be considered (see Table 2) and a summary of their effects is discussed below.

Table 1. Conventional product classification for micro-injection molding (µIM).

µIM Product	Average Part Size	Part Mass	Dimensional Tolerance Range	Equipment
Single Microparts	<10 mm	0.0001–0.1 g	10 µm	µIM metering and dosing system
Parts featuring micro- or nanostructures	>10 mm	> 0.1 g	0.01–1 µm (on features)	Conventional IM injection system
Micro precision IM Parts	>10 mm	> 0.1 g	10–100 µm	µIM and IM systems

Table 2. Modeling governing equations and aspects for injection molding (IM) simulation depending on scale size.

Macro/Meso	Micro (µ)	Nano (n)
Conservation of Mass	Wall-Slip Effect	
Conservation of Momentum	Surface Tension	
Conservation of Energy	Local HTC	Molecular Dynamics
Polymer constitutive equation (*pvT*)	Unvented air	
Viscosity model	Surface Roughness	

Regarding the discretization of the solution domain, it is well known that 3D elements are necessary for decomposing micro-components when the wall thickness to flow length ratio is no longer negligible [4–6]. In addition, a high fidelity calibration of injection pressure and flow length can be achieved when the feeding systems (gate, runner, sprue, and injection unit as a hot runner) are included in the simulation domain [7,8]. At the microscale, the no-slip boundary condition at the wall is no longer a valid hypothesis. In fact, Cao et al. showed that for microcavities, the required pressure drop to create the polymer filling induced polymer slippage at the walls. Thus, a non-zero-velocity boundary condition should be considered for the shear stress at the walls to solve the system of equations (Equations (1)–(3)) [9,10]. Moreover, capillary effects generated by surface tension forces become relevant, especially for nanoscale cavities. In order to include the surface tension and to account for the pressure loss/gain at the polymer melt interface, Rytka et al. [11] introduced an additional force in Equation (2). In the same study [11], polymer material properties such as flow temperature and contact angles were re-engineered or re-measured and modified in the simulation software database, since small variations of these parameters have been claimed to significantly affect the simulation results at the microscale. To calibrate the µIM simulations, another parameter that differs significantly from the macro scale simulations is the local heat transfer coefficient (HTC). Depending on the velocity gradient from the walls, different shear stress and viscosity are developed, leading to a different Nusselt number and HTC. Conventional simulation approaches use an average HTC for the filling and packing phases. However, since the thermal equilibrium between the polymer melt and mold establishes much faster at the walls than at the cavity core, a local HTC is assigned to the model properties when representing microfeatures. In this way, a more accurate prediction of the skin layer formation and of the flow rate can be found [9,12,13].

Other physical phenomena of central importance in µIM that need to be implemented in the simulation models are the effect of imperfect venting of micro-injection molds and the consequent counter pressure induced by the residual trapped air inside the tool's microcavity [14,15]. Furthermore, one other particularly challenging aspect for µIM process simulation is the implementation of the cavity's surface topography into the model. Even though this term was found to be significant in altering the filling behavior of microfeatures [16], the boundary layer on all the skin parts would require a consistent extension of the number of elements and time required for the simulation. In addition, it is not cost-effective to measure the entire surface roughness of a mold cavity and correctly model it in a digital form.

When focusing on parts featuring surface micro/nano structures, the implementation of simulation protocols becomes even more challenging due to the intrinsic multiscale nature of the domain. In order to reduce computational and modeling effort, the µIM simulation is often broken down in two separated steps: first, a macro/meso scale simulation of the part cavity without surface structures, which is followed by a second micro/nano simulation of the single surface feature where additional physics are added based on the size and geometry of the features themselves. In the first step, the temperature, pressure, and velocity of the polymer melt during injection are found and fed into the second step as the boundary conditions. For nanoscale features, the size of the polymer chains can also be considered using molecular dynamics simulation approaches [17–19].

This sequential method is not scalable because it requires case-by-case multi-step validation for each boundary condition that needs to be extracted, and is fundamentally based on the assumption that the boundary conditions are providing a sufficiently good approximation as the start input for the finer model. For this reason, an integrated multi-scale approach would be preferred. With the term multi-scale, a single simulation that combines a multi-scale mesh or multi-scale model formulation is intended. In Figure 1, a case selection of modeling studies shows the employed simulation method in comparison to the feature aspect ratio and its size [9–11,17–29].

This work proposes four case studies in which integrated multi-scale µIM process simulations are employed as a digital process optimization tool. In the first case, the calibration of the model using the effective mold microfeature dimension was used for the selection of process parameters in single micropart production. In the second case, a full factorial design of experiments and simulations investigated how to predict flash formation in a single micropart production by adding the venting channel as part of the cavity domain. In the third and fourth cases, meshing and domain partitioning strategies were proposed. The presented approaches aimed to realize a unique integrated multi-scale method for the investigation of parts featuring low-aspect-ratio microstructures for optical applications.

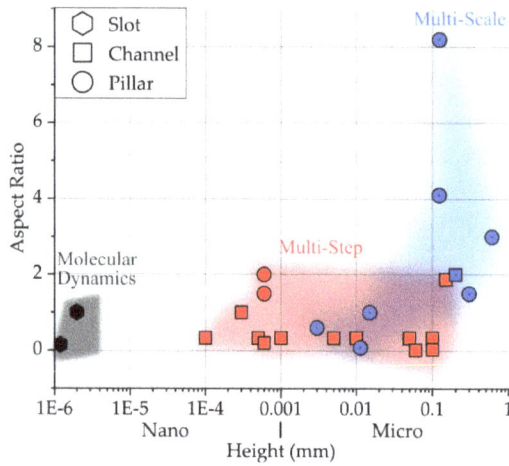

Figure 1. Comparison of μIM modeling cases for parts featuring micro/nano surface structures based on the structures' aspect ratio, height, and feature geometry [9–11,17–29].

2. Optimization of 3D Micropart Production

2.1. Case 1—Micro-Ring

The first case under analysis focused on the implementation of the μIM process simulation for the optimization of the production of a micro-ring. The part was a ring with a nominal outer diameter of 1.5 mm and an internal diameter of 0.45 mm. The part was manufactured with a commercially available thermoplastic elastomer (TPE) (Cawiton® PR 10589 F, B.V. Rubberfabriek Wittenburg, Zeewolde, the Netherlands) and its final mass was 2.2 mg. The manufacturing tolerance for the outer diameter production was 10 μm. The design of the part is reported in Figure 2. The purpose of the simulation was to predict the dimension of the outer diameter and find an optimal set of process parameters that would allow for production within the specifications.

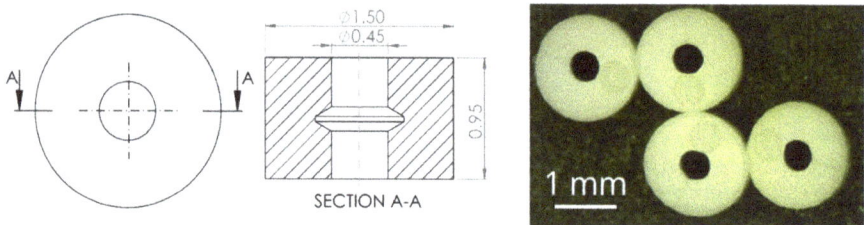

Figure 2. Design specification of the polymer micro-ring (**left**) and micro-molded parts (**right**) [30].

2.2. Case 2—Micro-Cap

The second case refers to another single micro-component that had a nominal part weight of 0.1 mg. The part is depicted in Figure 3, and has application in the medical industry. The part was made of a high-flowability commercially available polyoxymethylene (POM) (Hostaform C 27021, Celanese Corporation, Dallas, TX, USA) and had a hollow tapered internal geometry. The critical production aspect of the part consists of the formation of flash at the end of the part and μIM simulation was employed for the prediction and evaluation of flash formation based on a given set of process parameters.

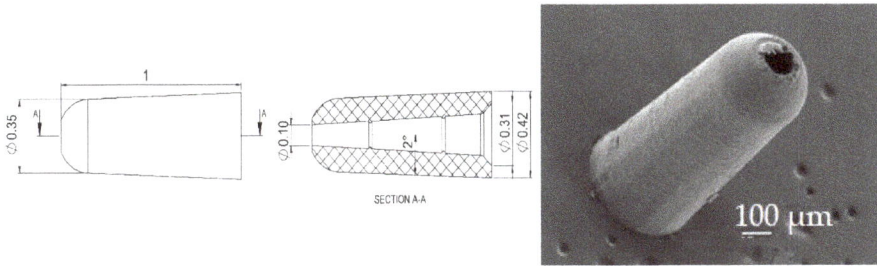

Figure 3. Design specifications of the polymer micro-cap in the analysis (**left** and **center**); scanning electron microscope image of the micro-injection molded part (**right**) [31].

2.3. Multi-Scale Modeling and Meshing of Single Micro-Components

Both simulations in Case 1 and Case 2 were implemented in Moldflow® Insight 2017 (Autodesk Inc., Melbourne, Australia) software using a multi-scale approach. No additional subroutines were developed in addition to the commercially available version of the software. The mesh was generated in both cases using tetrahedral elements with varying dimensions from 50 μm to 500 μm for Case 1 (see Figure 4) and 20 μm to 300 μm for Case 2 (see Figure 5). The entire feeding system was included in the models as it takes most of the overall material to produce single components. The total number of elements handled in the simulation was 1.4 million in Case 1 and 1.0 million in Case 2. The material properties in terms of pvT and viscosity follow a double-domain Tait model equation and a Cross-WLF model, both available in the software library. Experimental μIM parts were produced using a MicroPower 15 machine, (Wittmann Battenfeld Vienna, Austria). A summary table of the simulation multi-scale parameters is presented in Table 3.

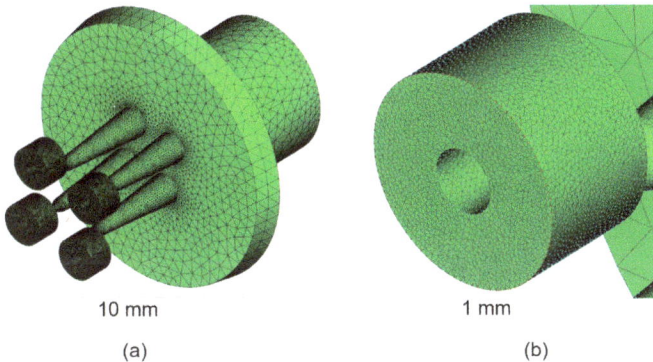

Figure 4. Multi-scale mesh of Case 1 including: sprue, runners, and multicavity tooling system (**left**); detailed view of the molded part and of the venting structure (**right**) [31].

In Case 1, the measured mold diameter was used as the nominal dimension. The outer diameter of the micro-ring cavity was measured using an optical coordinate measuring machine (DeMeet 3D, Schut Geometrical Metrology, Groningen, The Netherlands). An initial validation of the model was performed by comparing the diameters of the measured part and the simulated one. An initial deviation of 21 μm was measured from the experimental and simulated results. The model validation was performed using a parametric approach. The nominal outer diameter in the simulation was systematically reduced by 5 μm and the respective simulated result was compared to the experimental result until the final deviation from the values was found to be lower than the uncertainty of the measurement of 2 μm. The total number of iterations to achieve an attuned model was six, and the individual results are reported in Figure 6. A linear correlation was found from the nominal and resulting diameter, finding a uniform deviation from the simulation input and resulted in a calculated distance of (38 ± 3) µm. This finding indicates that the systematic difference could be amended by a correction factor (0.98 in this case), achieving an accuracy below the measurement uncertainty. It was assumed that a combination of factors including the actual dimension of the other component features as well as the effective shrinkage of the material at processing conditions, influenced the magnitude of the correction factor.

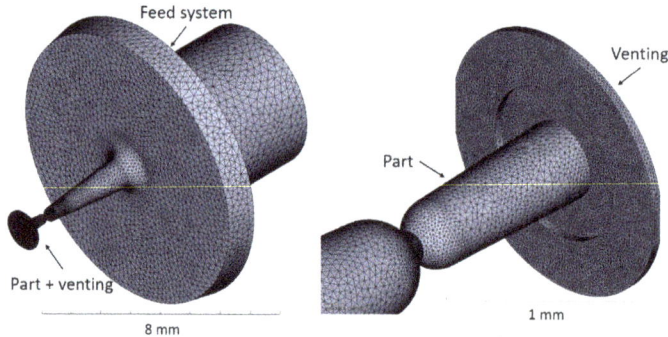

Figure 5. Multi-scale mesh and modified model geometry of Case 2 to include the flash formation in the simulation of the micro-cap [31].

Table 3. Multi-scale meshing parameters for the single micro-plastic components.

Mesh Parameters	Case 1 Micro-Ring	Case 2 Micro-Cap
Element type	3D Tetrahedral	3D Tetrahedral
Meshing algorithm	Advancing Front	Advancing Front
Modeling of sprue	Yes	Yes
Number of cavities	4	1
Minimum element size	50 µm	20 µm
Maximum element size	500 µm	300 µm
Growth rate	1.2	1.2
Total elements	1.4×10^6	1.0×10^6

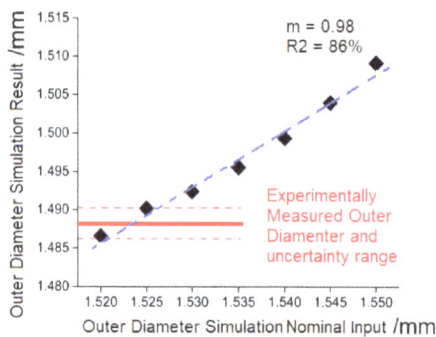

Figure 6. Design specification of the demonstrator with light selective reflection surface structures.

In Case 2, the model design was modified in order to include flash in the molded part. This was achieved by adding an outer ring that develops radially from the component right-end of Figure 3. The modified model of the part is shown in Figure 5. The thickness of the flash domain was set in two areas, the first one spreading up to twice the cap outer diameter to 0.84 mm with a thickness of 10 µm, while the second part spread to an outer diameter of 1.26 mm with a thickness equal to 20 µm. The definition of these geometrical properties was based on the outcome of different parametric trials.

2.4. Process Optimization Based on Calibrated Simulation Results—Case 1

For Case 1, the calibrated simulation model was subsequently used to run an optimization campaign based on four factors: mold temperature, melt temperature, injection velocity, and holding pressure. The temperature factors were varied on two levels, {30, 40} °C and {210, 225} °C, while velocity and pressure were varied on three levels, {50, 70, 90} mm/s and {30, 50, 70} MPa, respectively. The total number of the combinations was $2^2 \times 2^3 = 36$. The simulation results were compared with equivalent experimental analysis, which included five process replications and three repeated measurements for a total number of 540 data points. In Figure 7, the comparison of the main effects of each individual process factor on the outer diameter is reported.

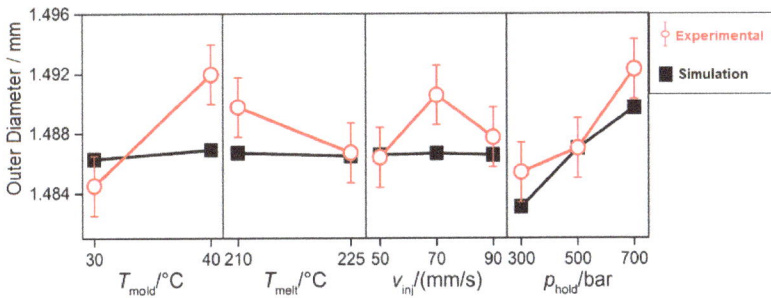

Figure 7. Main effects plot of the process factors on the outer diameter of the micro-ring [31].

The main effects showed an average deviation of (2.1 ± 2.0) µm from the simulated and experimental values with a maximum deviation of 5.5 µm observed for the cases at high mold temperature. The other cases indicated a deviation within 3 µm. These results validate the possibility of using the simulation as a calibrated tool for process optimization. As a matter of fact, an additional factorial design was simulated to find the optimal process conditions with a range of wider factor levels. There were three new factors in the analysis: mold temperature {50, 60} °C, melt temperature {180, 195} °C, and holding pressure {90, 110} MPa. The results of the investigation are shown in Figure 8, which also presents the target nominal diameter as well as the uncertainty of measurements and the tolerance limit. In this case, the µIM process simulation was able to predict that half the process conditions (those corresponding to high holding pressure, i.e., 110 MPa) investigated in the analysis would yield parts out of specification when the measurement uncertainty was considered.

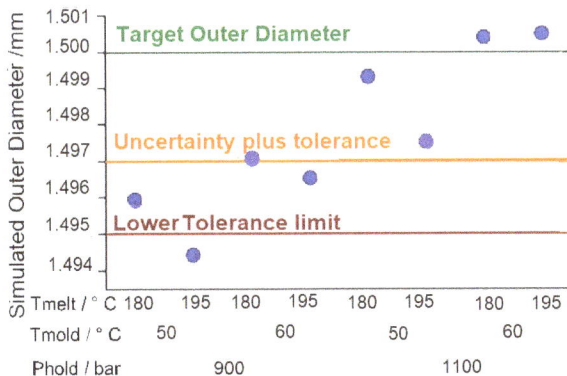

Figure 8. Simulated optimal process conditions for the outer diameter of the micro-ring.

2.5. Optimization of the Flash Formation Using Process Simulation—Case 2

For Case 2, a full factorial design was performed with the following factor levels: mold temperature (100–110 °C), melt temperature (200–220 °C), injection velocity (150–350 mm/s), and holding pressure (25–50 MPa). The total number of experiments was, in this case, replicated five times and the simulation was carried out once for each process condition. The simulated flash formation at the end of the cavity filling is shown as the result of the step-by-step flow simulations in Figure 9a. The area of the experimental flash (Figure 9b) and of the simulated flash (Figure 9c) was then measured by image processing for each process combination of both the real and virtual DOE. In Figure 10, the comparison of the main effects affecting the measured flash formed area against the simulated once is reported. The scale bars of the simulated and actual experiments ranged on a scale of 7 μm. Although the nominal values of the simulated results were overestimated by a factor 2.1, in relative terms, this means that the simulation predicted a flash area that was twice as large as the actual experimental case. Nonetheless, the effect amplitude and sign were congruent for injection velocity, melt temperature, and mold temperature variations, indicating that the amplitude value could be calibrated by the previously mentioned factor to obtain an accurate prediction of the flash area for a given set of molding parameters. With this result, it is possible to use μIM process simulation to find the effect of process parameters on flash formation and at the same time find a calibration factor for the design of the microcavity vent.

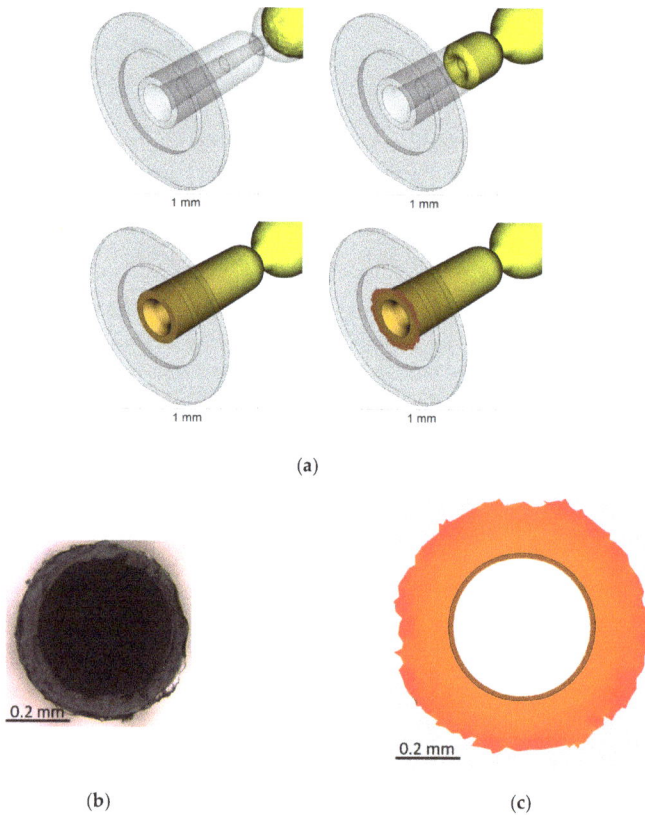

(a)

(b)   (c)

Figure 9. (**a**) Simulated flash formation during cavity filling; (**b**) real flash on the part; and (**c**) simulated flash (the dimensional scale bar is equal in both images) [31].

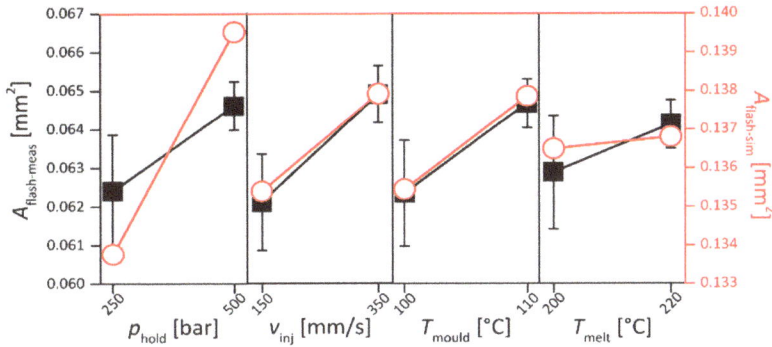

Figure 10. Main effects plot of the flash area for the experiments $A_{\text{flash-meas}}$ (in black) and simulations $A_{\text{flash-sim}}$ (in red). Note that the scales for the two sets of results are different, but the ranges shown are equal. Interval bars represent the standard errors of experimental data [31].

3. Multi-Scale Filling Simulation of Low Aspect Ratio Structures in Parts Embedding Microfeatures

3.1. Case 3—Micro-Optical Reflector

The use of µIM process simulation was further extended to parts embedding a microstructured surface with low aspect ratio micro-grooves. The Case 3 component is an optical demonstrator with surface features that enable light selective reflection (Figure 11). The structures consist of parallel triangular grooves with a nominal height of 34 µm, a width of 200 µm, and a slope of 10°, resulting in a growing aspect ratio from 0 to 0.17. The part was molded in a commercially available acrylonitrile butadiene styrene (ABS) (Terluran® GP35, INEOS Styrolution Group GmbH, Frankfurt, Germany) and the mass of the actual part was 401 mg.

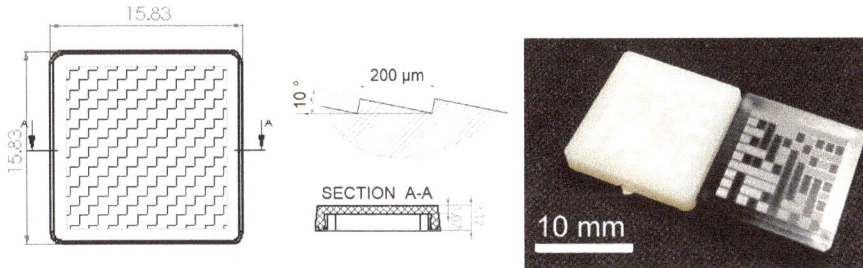

Figure 11. Design specification of the micro-optical reflector (**left**) with a highlight of its surface structures (**center**); injection molded part and tool cavity insert (**right**) [32].

3.2. Case 4—Fresnel Lens

Case 4 is represented by a Fresnel lens whose surface is structured by low aspect ratio features. The specimen was manufactured in a commercially available cyclo-olefin polymer (COP) (Zeonex E48R®, Zeon Corporation, Tokyo, Japan) and had a total mass of 13.4 g. The part had global dimensions of 60 mm × 85 mm and was covered with concentric surface microgrooves. The structures were semi-pyramidal with a constant pitch of 749 µm and a varying height from 17 µm to 346 µm with a growing aspect ratio of 0.02 to 0.46 from the center to the outer of the structure array. The total array covered an area of 40 mm × 40 mm and is shown in Figure 12.

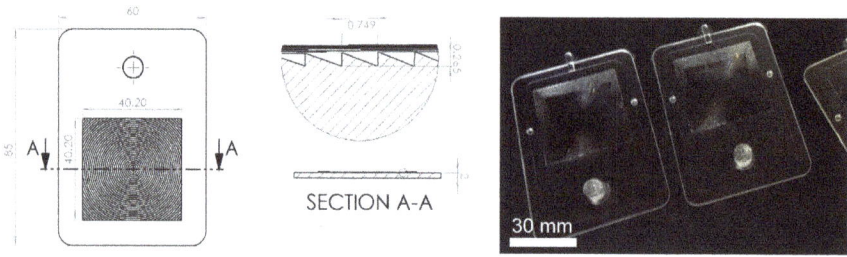

Figure 12. Design specification of the analyzed Fresnel lens (**left**) and detail of the low aspect ratio surface structures (**center**); injection molded parts (**right**) [33].

3.3. Multi-Scale Meshing for Microstructured Parts

Both microstructured parts exhibited a highly multi-scale nature of the design with a minimum to maximum feature dimension ratio of 34 μm (microfeature height)/15.830 mm (part edge length) (i.e., aspect ratio 1/466) for the reflector and 17 μm (minimum microfeature height)/85.000 mm (part long edge length) (i.e., aspect ratio 1/5000) for the Fresnel lens. When generating a mesh that handles this aspect ratio, it is necessary to find a compromise between the meshing parameters (minimum/maximum element size, growth rate, and aspect ratio of each element), computational time, and accuracy. Some of the previously mentioned software providers allow for hybrid mesh generation that refines the element size at the surface in order to address the surface phenomenon and features. Others allow for mesh partitioning in order to generate a local surface refinement where required. For the two proposed cases, the Autodesk Moldflow® Insight 2019 (Autodesk Inc., Melbourne, Australia) software was employed for mesh generation and simulation. No additional subroutines were developed in addition to the commercially available version of the software. 3D tetrahedral elements were employed. Two different approaches were employed in order to find a compromise between the computational effort and the simulation accuracy.

For Case 3, only a fraction of the microfeatures array was added to the simulation domain (i.e., in the simulated geometry). The features closer to the gate were included due to the lower replicability that was observed experimentally [32]. An advancing layer [34] algorithm was used to first generate the surface mesh elements, and subsequently in the mesh elements across the part thickness. The meshing parameters are summarized in Table 4. The part, in this case, was produced in a four cavity mold and the resulting mesh is shown in Figure 13.

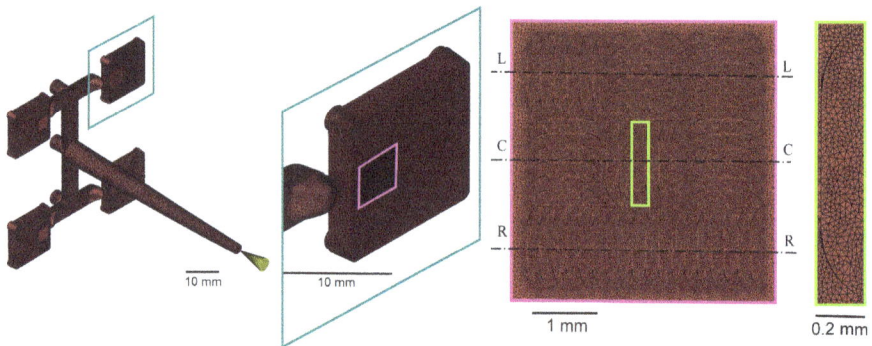

Figure 13. The multi-scale mesh of the optical demonstrator with selected surface features included in the geometry [35].

For Case 4, the whole surface microfeatures were included in the modeled part geometry, but only a part of them was locally refined using surface mesh partitioning areas. The mesh was created using an advancing front [36] algorithm by first generating the surface elements. A local refinement on the surface microfeatures down to an element size of 10 µm was performed, as shown in Figure 14. Additional meshing parameters are shown in Table 4.

Table 4. Multi-scale meshing parameters for the micro-structured plastic parts.

Mesh Parameters	Case 3 Micro-Optical Reflector	Case 4 Fresnel Lens
Element type	3D Tetrahedral	3D Tetrahedral
Meshing algorithm	Advancing Layer	Advancing Front
Meshing approach	Feature restriction	Partition refinement
Minimum element size	10 µm	10 µm
Maximum element size	1.000 mm	1.000 mm
Growth rate	1.2	1.5
Total elements	3 047 407	3 248 186

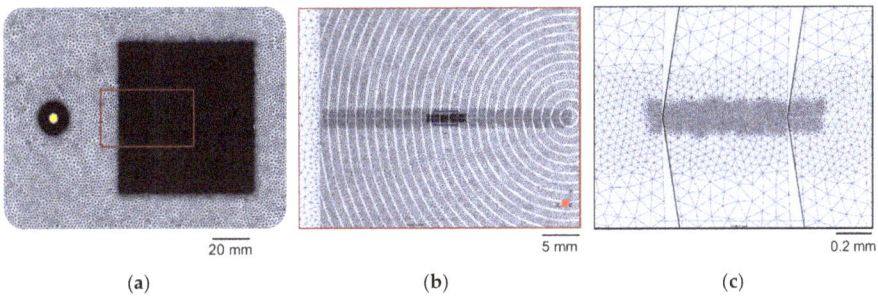

20 mm	5 mm	0.2 mm
(a)	(b)	(c)

Figure 14. The multi-scale mesh of the Fresnel lens with partial surface mesh refinement at (a) part level, (b) structured area level, and (c) at single micro feature level.

3.4. Multi-Scale Filling Simulation Validation at Mesoscale—Cases 3 and 4

Validation of the filling behavior requires an experimental comparison of injection molded parts with the simulation results. For Case 3, an Allrounder 370A injection molding machine from Arburg (Loßburg, Germany) was employed with an injection screw of 18 mm in diameter. For Case 4, the employed machine was a Ve70 from Negri Bossi (Cologno Monzese, Italy) equipped with an injection screw with a diameter of 26 mm. The injection screw absolute velocity over time and its initial position were used as input parameters for the injection molding simulation. The profiles are reported in Figure 15a for Case 3 and Figure 15b for Case 4.

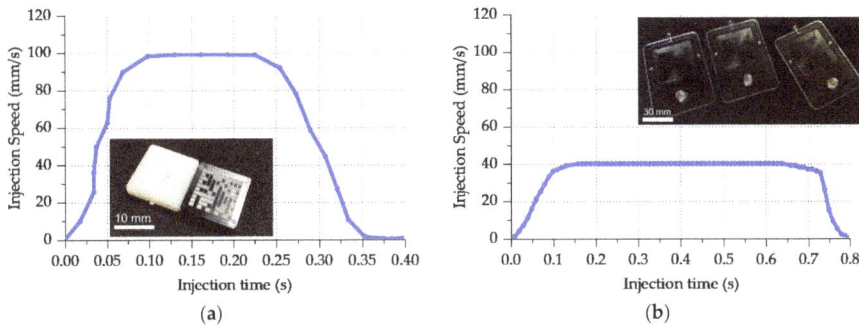

(a)   (b)

Figure 15. Injection speed profiles sampled from the machine controller for Cases 3 (a) and 4 (b).

The simulation validation for Cases 3 and 4 included the injection pressure profiles, as shown in Figure 16. The actual value was extracted directly from the injection molding machine control interface and represents the pressure in the filling phase. The validation plot shows that during filling, the actual simulated pressure required to fill the cavity was higher than the simulation. For Case 3, the simulated integral over time of the injection pressure was 20.1 MPa s against 22.0 MPa s with an underestimation of pressure over time by 1.9 MPa s, which in this case was 8% of the actual value. For Case 4, the simulated integral had a value of 40.1 MPa s against the effective 41.5 MPa s with a nominal deviation of 1.4 MPa s, which in relative terms was 3% of the actual value. The pressure underestimation was attributed to the presence of surface roughness and the absence of a full multi-scale surface microfeature domain.

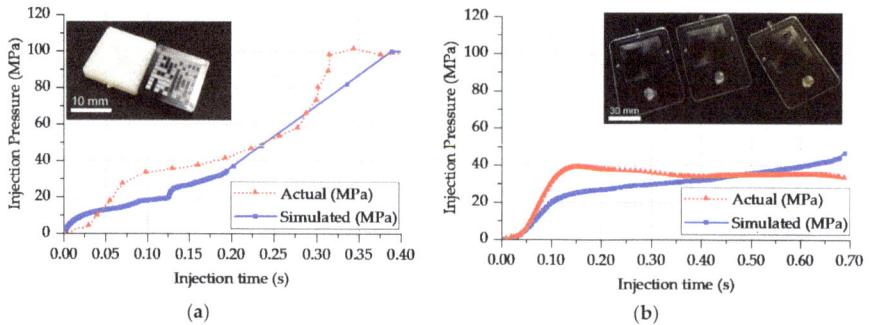

Figure 16. Injection pressure at the injection location comparison from the simulation and actual value for Cases 3 (**a**) and 4 (**b**).

3.5. Multi-Scale Simulation for Filling Time Prediction—Case 3

At the microscale, the surface features were fully replicated in both of the considered design solutions. Specifically for Case 3, the restricted features under analysis were compared in terms of filling time. A total of 250 intermediate results during filling were calculated to achieve a time resolution of 0.8 ms. A filling time comparison from the top edge of the microfeature and corresponding bottom polished surface at the same X coordinate was made for the three sections shown in Figure 13 intended as left (L), central (C), and right (R). The results of filling time as function of the X coordinate (i.e., the flow length) are reported in Figure 17a–c. A comparison of residual filling time is then presented in Figure 17d. As can be seen in Figures 12a and 17b, the orientation of the microstructures is orthogonal with respect to the flow propagation direction (i.e., the flow front velocity vector main direction). As a matter of fact, it can be seen that the melt flow was delayed when filling of the microfeatures occurred (Figure 17a,b) in comparison to when the flow was filling the bottom flat (i.e., unstructured, surface, see Figure 17c).

The delay followed a quadratic trend and was found for the two different and opposite sections of the flow front. In contrast, at the center, the orientation of the microfeatures is oriented longitudinally to the melt front, and as shown in Figure 17c, there is no induced delay of the filling time. The result indicates that μIM simulation can be used to predict the influence of surface microstructure arrays on melt front propagation. This result is an outcome of the continuity and conservation of linear momentum, and differently from other studies [37–39], neglects the wall slip. This result leads to the conclusion that surface microfeature arrays and their orientation affects the melt front propagation; leading to an even greater need for fully-integrated multi-scale models to avoid inaccurate estimation of velocity for a multi-step modeling approach. The mesh with a restricted portion of surface features allowed finding a correlation between flow length and filling time of periodic microfeatures.

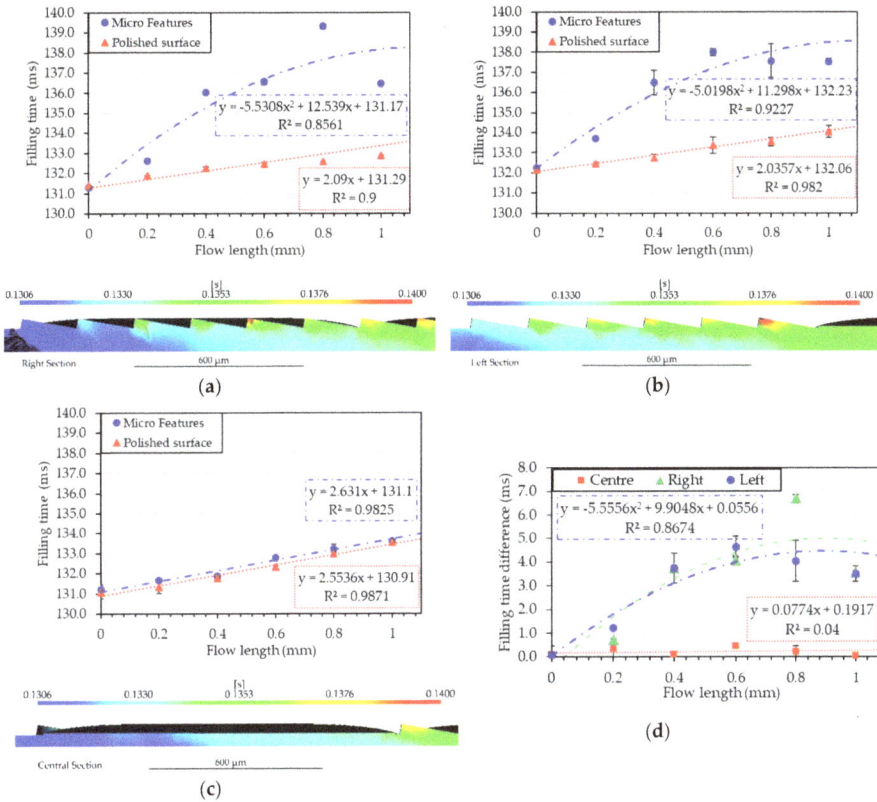

Figure 17. Filling time comparison from the microfeature and bottom polished surface as a function of the flow length for the right (**a**), left (**b**), and central (**c**) section of Case 3 with the summary difference of filling time for each condition (**d**).

3.6. Multi-Scale Simulations for Microfeature Replication Prediction—Case 4

In Case 4, the multi-scale approach was employed for the evaluation of microfeature replication during the molding process. The task was solved by performing injection molding of a short shot that corresponded to a condition where a microfeature was partially replicated. The aim of this process was to show a methodology that could be used to validate the simulation with the experimental data. The evaluation of microfeatures was done experimentally by measuring shots using a laser confocal microscope (LEXT OLS4100, Olympus, Tokyo, Japan). A cross-section of the sample image was exported and considered as the reference profile. In a virtual environment, the simulated short shot was measured after the extraction of the surface elements coordinates and sectioning along the same axis. To do so, the filling results were exported in the (*.stl) format, and the surface nodal vertex belonging to the cross-section in the medial axis of the specimen was sampled and plotted to find the simulated profile of the microfeature. Two-time instances were considered for the comparison: an initial case when half the microfeature was filled corresponding to a short shot, as shown in Figure 18a,c, and for full replication, as shown in Figure 18b,d. The shot was sampled at the same injection screw displacement (16 mm) at a simulated time of 0.485 s. The melt front propagation at the mesoscale was compared between the simulation and the experiments by overlapping the top view of the simulated filling step with an optical image of the parts obtained by stitching individual confocal microscope color image acquisitions. For the full replicated part, the same validation at the mesoscale was conducted. Parts were analyzed comparing the flow front final length and injection stroke (28 mm) at an injection time of 0.752 s. At the microscale, the feature on the short shots had different height values in terms of peak-to-valley measurements. The simulated feature height was 79 ± 10 µm and the measurements on the molded feature led to 56.9 ± 4.7 µm. The simulated value for the full replication step was equal to the nominal dimension 185.0 ± 10 µm (i.e., 100% replication), while the actual value was in fact 169.0 ± 4.7 µm (i.e., 91% replication). The tip radius of the measured feature (55 µm) was a combination of both the actual mold finite edge, which for the metrological limitation and implementation on the full scale of the simulation model were not included, and the replication factor of the molding process.

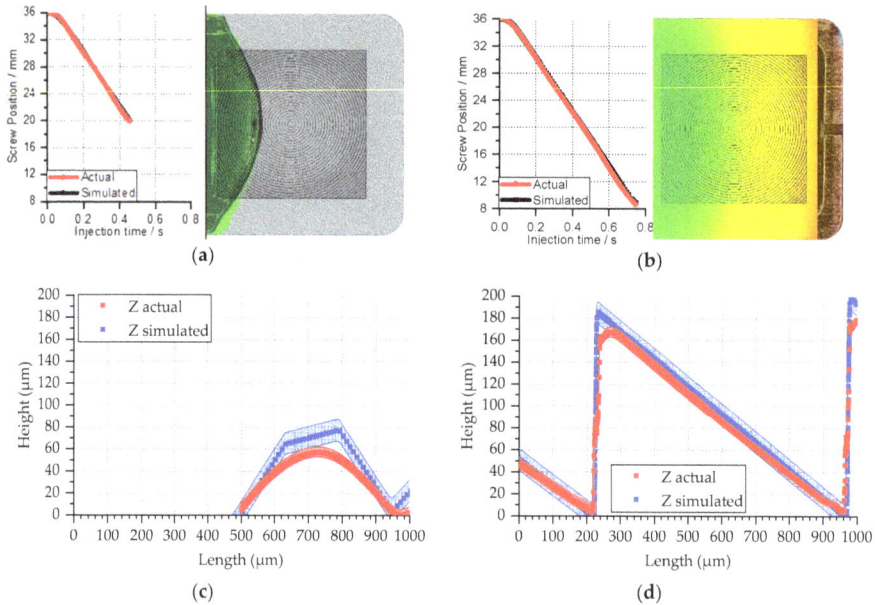

Figure 18. Filling volume comparison from the actual and simulated values for a short shot at meso- (**a**) and microscale (**c**), and full replication for the meso- (**b**) and microscale (**d**).

4. Conclusions

The current work has shown the current state-of-the-art in the use of µIM simulation capabilities. In the study, the progress of multi-scale methods against multi-step approaches was further explained by use cases for single three-dimensional microparts and parts embedding surface microstructures. From the experimental analysis and the corresponding simulation validation conducted on these cases, the following conclusions can be obtained:

- Process simulation requires geometrical calibration of the domain by measuring the effective feature size on the mold insert and feeding this value to the simulation boundary condition.
- µIM process simulation can be used for the optimization of single-part production with a 1 mm feature dimension at a 10 µm accuracy level.
- µIM process simulation can be used in the prediction of the factors most affecting flash formation in single micropart production. The punctual estimation of the flash area requires further calibration of the model geometry and a venting flow volume has to be included in the part design.
- Virtual design of experiments using simulations is an effective digital optimization tool that has the capability to indicate the effect of µIM process parameters on micro-molded part characteristics.

- For parts featuring microfeatures, two methods were proposed for the modeling of complex microstructure arrays: (1) feature restriction and (2) feature refinement.
- The feature restriction approach allowed us to model the filling time delay from the flat polished and microstructured side of the cavity, allowing us to predict the influence of feature orientation on the melt front propagation.
- Feature refinement allowed us to punctually investigate the replication development of a single microfeature. Through combined meso- and micro-dimensional scale comparison, a multi-scale validation approach was proposed.

Author Contributions: Conceptualization, D.L., F.R., F.B., M.C., D.Q., Y.Z., and G.T.; Data curation, D.L., F.R. and F.B.; Formal analysis, D.L., F.R., and F.B.; Investigation, D.L., F.R., F.B., and G.T.; Methodology, D.L., F.R., F.B., M.C., and G.T.; Software, D.L., F.R., and F.B.; Visualization, D.L., F.R., and F.B.; Funding acquisition, M.C., Y.Z., and G.T.; Resources, M.C., D.Q., Y.Z., and G.T.; Supervision, M.C., D.Q., Y.Z., and G.T.; Project administration, M.C., D.Q., Y.Z., and G.T.; Writing—original draft preparation, D.L. and G.T.; Writing—review and editing, D.L., F.R., F.B., M.C., D.Q., Y.Z., and G.T. All authors have read and agreed to the published version of the manuscript.

Funding: This research work was undertaken in the framework of the following projects: MICROMAN ("Process Fingerprint for Zero-defect Net-shape MICROMANufacturing", http://www.microman.mek.dtu.dk/), Marie Skłodowska-Curie Action Innovative Training Network funded by Horizon 2020; the European Union Framework Program for Research and Innovation (Project ID: 674801), ProSurf ("High Precision Process Chains for the Mass Production of Functional Structured Surfaces", http://www.prosurf-project.eu/), funded by Horizon2020, the European Union Framework Program for Research and Innovation (Project ID: 767589); MADE ("Manufacturing Academy of Denmark") DIGITAL (https://en.made.dk/digital/), WP3 "Digital Manufacturing Processes", funded by Innovation Fund Denmark (Project ID: 6151-00006B); QRprod ("QR coding in high-speed production of plastic products and medical tablets"), funded by Innovation Fund Denmark (Project ID: 6151-00006B).

Acknowledgments: Contributions to this work were provided by Úlfar Arinbjarnar, Alberto Garcia Martinez, Fililppo Degli Esposti, Frederik Boris Hasnæs, students of the Department of Mechanical Engineering at the Technical University of Denmark.

Conflicts of Interest: The authors declare no conflicts of interest.

References

1. Tosello, G. *Micro Injection Molding*, 1st ed.; Carl Hanser Verlag: Munich, Germany, 2018.
2. Kennedy, P. *Flow Analysis of Injection Molds*, 2nd ed.; Carl Hanser Verlag: Munich, Germany, 2013; pp. 188–190.
3. Chang, H.H.; Hieber, C.A.; Wang, K.K. A unified simulation of the filling and post-filling stages in injection molding: I Formulation. *Polym. Eng. Sci.* **1991**, *31*, 116–124. [CrossRef]
4. Kim, S.W.; Turng, L.S. Developments of three-dimensional computer-aided engineering simulation for injection moulding. *Model. Simul. Mater. Sci. Eng.* **2004**, *12*, 151–173. [CrossRef]
5. Kim, S.W.; Turng, L.S. Three-dimensional numerical simulation of injection molding filling of optical lens and multi-scale geometry using finite element method. *Polym. Eng. Sci.* **2006**, *46*, 1263–1274. [CrossRef]
6. Yu, L.; Lee, L.J.; Koelling, K.W. Flow and heat transfer simulation of injection molding with microstructures. *Polym. Eng. Sci.* **2004**, *44*, 1866–1876. [CrossRef]
7. Tosello, G.; Costa, F.S. High precision validation of micro injection molding process simulations. *J. Manuf. Process.* **2019**, *48*, 236–248. [CrossRef]
8. Guerrier, P.; Tosello, G.; Hattel, J.H. Flow visualization and simulation of the filling process during injection molding. *CIRP J. Manuf. Sci. Technol.* **2017**, *16*, 12–20. [CrossRef]
9. Cao, W.; Kong, L.; Li, Q.; Ying, J.; Shen, C. Model and simulation for melt flow in micro-injection molding based on the PTT model. *Model. Simul. Mater. Sci. Eng.* **2011**, *19*, 085003. [CrossRef]
10. Rajhi, A.A.; Jedlicka, S.S.; Coulter, J.P. Moldflow optimization of micro-cavities filling during injection molding process. In Proceedings of the ANTEC International Conference, Orlando, FL, USA, 7–10 May 2018.
11. Rytka, C.; Lungershausen, J.; Kristiansen, P.M.; Neyer, A. 3D filling simulation of micro- and nanostructures in comparison to iso- and variothermal injection moulding trials. *J. Micromech. Microeng.* **2016**, *26*, 065018. [CrossRef]
12. Babenko, M.; Sweeney, J.; Petkov, P.; Lacan, F.; Bigot, S.; Whiteside, B. Evaluation of heat transfer at the cavity-polymer interface in microinjection moulding based on experimental and simulation study. *Appl. Therm. Eng.* **2018**, *130*, 865–876. [CrossRef]

13. Cui, Z.X.; Si, J.H.; Liu, C.T.; Shen, C.Y. Flowing simulation of injection molded parts with micro-channel. *Appl. Math. Mech.* **2014**, *35*, 269–276. [CrossRef]

14. Griffiths, C.A.; Dimov, S.S.; Scholz, S.; Tosello, G. Cavity Air Flow Behavior During Filling in Microinjection Molding ASME. *J. Manuf. Sci. Eng.* **2011**, *133*, 11006. [CrossRef]

15. Sorgato, M.; Masato, D.; Lucchetta, G. Effect of vacuum venting and mold wettability on the replication of micro-structured surfaces. *Microsyst. Technol.* **2017**, *23*, 2543–2552. [CrossRef]

16. Surace, R.; Sorgato, M.; Bellantone, V.; Modica, F.; Lucchetta, G.; Fassi, I. Effect of cavity surface roughness and wettability on the filling flow in micro injection molding. *J. Manuf. Process.* **2019**, *43*, 105–111. [CrossRef]

17. Zhou, M.; Jiang, B.; Weng, C. Molecular dynamics study on polymer filling into nano-cavity by injection molding. *Comput. Mater. Sci.* **2016**, *120*, 36–42. [CrossRef]

18. Li, Y.; Abberton, B.C.; Kröger, M.; Liu, W.K. Challenges in multiscale modeling of polymer dynamics. *Polymers* **2013**, *5*, 751–832. [CrossRef]

19. Zhang, M.; Xin, Y. Molecular mechanism research into the replication capability of nanostructures based on rapid heat cycle molding. *Appl. Sci.* **2019**, *9*, 1683. [CrossRef]

20. Choi, S.J.; Kim, S.K. Multi-scale filling simulation of micro-injection molding process. *J. Mech. Sci. Technol.* **2011**, *25*, 117–124. [CrossRef]

21. Lee, W.L.; Wang, D.; Wu, J.; Ge, Q.; Low, H.Y. Injection Molding of Superhydrophobic Submicrometer Surface Topography on Macroscopically Curved Objects: Experimental and Simulation Studies. *Appl. Polym. Mater.* **2019**, *1*, 1547–1558. [CrossRef]

22. Tofteberg, T.R.; Andreassen, E. Multiscale Simulation of Injection Molding of Parts with Low Aspect Ratio Microfeatures. *Int. Polym. Process.* **2010**, *25*, 63–74. [CrossRef]

23. Lin, H.Y.; Chang, C.H.; Young, W. Bin Experimental and analytical study on filling of nano structures in micro injection molding. *Int. Commun. Heat Mass Transf.* **2010**, *37*, 1477–1486. [CrossRef]

24. Kuhn, S.; Burr, A.; Kübler, M.; Deckert, M.; Bleesen, C. Study on the replication quality of micro-structures in the injection molding process with dynamical tool tempering systems. *Microsyst. Technol.* **2010**, *16*, 1787–1801. [CrossRef]

25. Shen, Y.K.; Chang, C.Y.; Shen, Y.S.; Hsu, S.C.; Wu, M.W. Analysis for microstructure of microlens arrays on micro-injection molding by numerical simulation. *Int. Commun. Heat Mass Transf.* **2008**, *35*, 723–727. [CrossRef]

26. Eom, H.; Park, K. Integrated numerical analysis to evaluate replication characteristics of micro channels in a locally heated mold by selective induction. *Int. J. Precis. Eng. Manuf.* **2011**, *12*, 53–60. [CrossRef]

27. Hong, J.; Kim, S.K.; Cho, Y.H. Flow and solidification of semi-crystalline polymer during micro-injection molding. *Int. J. Heat Mass Transf.* **2020**, *153*, 119576. [CrossRef]

28. Marhöfer, D.M.; Tosello, G.; Islam, A.; Hansen, H.N. Gate Design in Injection Molding of Microfluidic Components Using Process Simulations. *J. Micro Nano-Manuf.* **2016**, *4*, 025001. [CrossRef]

29. Dong, P.; Zhao, Z.; Wu, D.; Zhang, Y.; Zhuang, J. Simulation of injection molding of ultra-thin light guide plate with hemispherical microstructures. *Key Eng. Mater.* **2012**, *503*, 222–226. [CrossRef]

30. Baruffi, F.; Calaon, M.; Tosello, G. Effects of micro-injection moulding process parameters on accuracy and precision of thermoplastic elastomer micro rings. *Precis. Eng.* **2018**, *51*, 353–361. [CrossRef]

31. Baruffi, F. Integrated Micro Product/Process Quality Assurance in Micro Injection Moulding Production. Ph.D. Thesis, DTU Mechanical Engineering, Kongens Lyngby, Denmark, 2019.

32. Regi, F.; Doest, M.; Loaldi, D.; Li, D.; Frisvad, J.R.; Tosello, G.; Zhang, Y. Functionality characterization of injection moulded micro-structured surfaces. *Precis. Eng.* **2019**, *60*, 594–601. [CrossRef]

33. Loaldi, D.; Quagliotti, D.; Calaon, M.; Parenti, P.; Annoni, M.; Tosello, G. Manufacturing Signatures of Injection Molding and Injection Compression Molding for Micro-Structured Polymer Fresnel Lens Production. *Micromachines* **2018**, *9*, 653. [CrossRef]

34. Pirzadeh, S. Three-dimensional unstructured viscous grids by the advancing-layers method. *AIAA J.* **1996**, *34*, 43–49. [CrossRef]

35. Regi, F. Process Technologies for Functional Anisotropic Surfaces Generation in Quick Response Code Application. Ph.D. Thesis, DTU Mechanical Engineering, Kongens Lyngby, Denmark, 2019.

36. Löhner, R.; Parikh, P. Generation of three-dimensional unstructured grids by the advancing-front method. *Int. J. Numer. Methods Fluids* **1988**, *8*, 1135–1149. [CrossRef]

37. Orazi, L.; Sorgato, M.; Piccolo, L.; Masato, D.; Lucchetta, G. Generation and Characterization of Laser Induced Periodic Surface Structures on Plastic Injection Molds. *Lasers Manuf. Mater. Process.* **2020**, *7*, 207–221. [CrossRef]

38. Ebrahimi, M.; Konaganti, V.K.; Moradi, S.; Doufas, A.K.; Hatzikiriakos, S.G. Slip of polymer melts over micro/nano-patterned metallic surfaces. *Soft Matter* **2016**, *12*, 9759–9768. [CrossRef] [PubMed]

39. Lucchetta, G.; Masato, D.; Sorgato, M.; Crema, L.; Savio, E. Effects of different mould coatings on polymer filling flow in thin-wall injection moulding. *CIRP Ann. Manuf. Technol.* **2016**, *65*, 537–540. [CrossRef]

micromachines

MDPI

Article

Optimization of Nozzle Inclination and Process Parameters in Air-Shielding Electrochemical Micromachining

Minghuan Wang * , **Yongchao Shang, Kailei He, Xuefeng Xu and Guoda Chen**

Key Laboratory of Special Purpose Equipment and Advanced Processing Technology, Ministry of Education and Zhejiang Province, Zhejiang University of Technology, Hangzhou 310014, China; Shangyc@zjut.edu.cn (Y.S.); 2111802026@zjut.edu.cn (K.H.); xuxuefeng@zjut.edu.cn (X.X.); gchen@zjut.edu.cn (G.C.)

* Correspondence: wangmh@zjut.edu.cn; Tel.: +86-571-8529-0477

Received: 27 October 2019; Accepted: 3 December 2019; Published: 4 December 2019

Abstract: Microstructures on metal surfaces with diameters of tens to hundreds of micrometers and depths of several micrometers to tens of micrometers can improve the performance of engineering parts. Air-shielding electrochemical micromachining (AS-EMM) is a promising method for fabricating these microstructures, owing to its advantage of high efficient and better localization. However, the machining performance is often influenced by the machining or nonmachining parameters in AS-EMM. In order to get a better machining result in AS-EMM, the optimization of AS-EMM, including nozzle inclination and process parameters, was studied in this paper. Firstly, nozzle inclination was optimized by the different selected air incidence angles (θ) in simulation, and $\theta = \pi/4$ was advised. Then, the grey relational analysis based on the orthogonal test method was used to analyze the grey relational grade for parameters and obtain the optimal parameter combination, i.e., at electrolyte velocity 5.5 m/s, gas velocity 160 m/s, and voltage 8 V. Finally, the optimization result was verified experimentally.

Keywords: air-shielding electrochemical micro-machining; nozzle design; flow field; grey relational analysis; optimization

1. Introduction

Modern tribology [1,2], aerodynamics [3], and bionics [4] all confirm that metal surfaces with morphologies comprising microstructures of tens to hundreds of microns across can significantly improve equipment performance; for example, reducing friction, wear, and frictional resistance, improving stiffness, and via the removal of undesirable surface adhesion properties and seizures. Devaiah et al. [5] investigated the effect of a micropit on cutting performance of tools and found that it reduced the cutting force of tools and changed chip shape. Li et al. [6] filled the micropits with MnS_2 powder with different incidence rates, and found that the fraction coefficient decreased from 0.18 to 0.1, and the wear decreased. Xiao et al. [7] mentioned that dimples in mechanical seals could enhance heat transfer by increasing the solid–fluid contact area and mixing, thus reducing the seal's interface temperature by about 10%. Djamaï [8] found that thermo-hydrodynamic mechanical-face seals, which were equipped with a notched rotating face, were efficient to reduce friction. Zhuang et al. [9] studied microstructure heat exchangers with different forms of microstructures and different aspect ratios and found that the heat transfer performance improved with the increase of aspect ratio of microstructures.

Much attention has been paid to the manufacturing of textures on metals, and Electrochemical machining (ECM) is one of the nontraditional methods to machine materials into complex shapes with high precision. In ECM, the material was removed by electrolysis. Compared with other microstructure machining methods, ECM has advantages of no mechanical or thermal stress, no burrs or distortion of features [10], high surface quality, and high material removal rate (MRR), regardless of material hardness. It has many applications, one of which is electrochemical jet machining (EJM). EJM can control an electrolyte jet with a high velocity, concentrating on the local range for ejecting the electrolyte through a nozzle. Therefore, only the workpiece material exposed to the jet is removed by anodic dissolution. It is widely used in the machining of microstructures due to its high localization, large aspect ratio of machining morphology, and having no special requirements on tool shape. Liu et al. [11] investigated abrasive composite EJM by adding a solid micro-abrasive into the electrolyte. The anode dissolution and micro-abrasive erosion worked together to remove materials and improve the surface texture processing efficiency. Kunieda et al. [12] proposed an EJM application using a flat electrolyte jet, which offered a new idea for micromilling and electrochemical turning. Schubert et al. [13] applied superimposing multidimensional motions to a fabricated 3D complex structure with spiral geometry by EJM. Clare et al. [14] studied the influence of different parameters (jet angle, electrolyte, and current density) on the machining morphology by using a nozzle that could change the jet angle and produced complex structures with better precision. Hackert et al. [15] established a fluid dynamic model with COMSOL Multiphysics to simulate the formation and change of the liquid beam, and demonstrated the material removal process in EJM. Zhao et al. [16] established a multiphysics model to simulate EJM on inclined planes and studied the distribution of current density under different jet angles and workpiece configurations, and experimental verification was conducted to confirm the analysis. To reduce the undercutting, Chen et al. [17] proposed a method of conductive mask EJM for generating a micro-dimple. Speidel et al. [18] investigated different electrolytes for the purpose of enhancing the effect of processing. However, in EJM, severe stray corrosion around the machining area with opposite electrodes, which is caused by the hydraulic jump, would seriously affect the accuracy of microstructures. To solve this problem, Wang et al. [19,20] proposed a new processing method of air-shielding electrochemical micromachining (AS-EMM), which made a layer of compressed gas film coat the liquid beam coaxially and control the electrolyte to focus precisely on the workpiece area where the tool electrode was directly opposite.

Previous studies [19,20] have proved that AS-EMM can significantly reduce stray corrosion and improve the surface processing quality. However, the optimization of the nozzle inclination and machining parameters affecting the machining performance is essential. In this study, the nozzle was firstly optimized in simulation, based on the different incidence angles. Then, Grey relational analysis based on the orthogonal test method was used to develop a mathematical model for experimental results, and the built model was used as the objective functions for the multi-objective optimization. The process parameters were optimized to get the better performance of surface roughness and the micro pit's aspect ratio. Finally, the optimization result was verified experimentally.

2. Schematic of the Air-Shielding Electrochemical Micromachining (AS-EMM)

The schematic of AS-EMM is shown in Figure 1. The experimental configuration includes the machining tool, the control system (computer), the electrolyte recycle system, an air pump, power supply, and a nozzle. The air pump and electrolyte recycle system provide gas and electrolyte in the interelectrode gap (IEG) with a certain amount of pressure, respectively. The machine tool is controlled by the control system. The cathode is fixed in the spindle of the machine tool through the nozzle. Material on the workpiece surface is eroded via the power supply when connected.

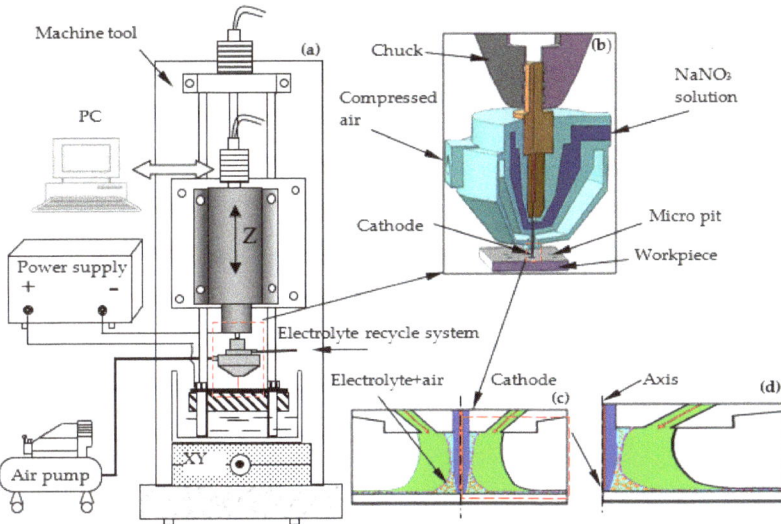

Figure 1. Schematic of (**a**) air-shielding electrochemical micromachining (AS-EMM) experimental set-up; (**b**) nozzle structure; (**c**) the flow field in interelectrode gap; (**d**) research zone.

3. Optimization of Air Incidence Angle

3.1. Model Description

In AS-EMM, nozzle inclination influences the flow field in the machining gap and then the machined micro pits. The optimization of the nozzle was done in simulation. The numerical model was created, and Fluent was employed to analyze the flow field characters. Due to the axis symmetry, the model was built as in Figure 2. There were 30,854 triangle units in total in this model. Considering the electrolyte and air in the machining gap, the mixture model was selected. The inlets' boundary condition for air and the electrolyte were 0.15 MPa and 0.1 MPa, respectively. The initial outlet pressure was 0 MPa. The distance from the outlet of electrolyte to the workpiece surface is 5 mm. The effect of the compressed air incidence angle on air void fraction and velocity in the flow field was investigated. In AS-EMM, lower air fraction in the machining area can improve conductivity of flow, so that the MRR is improved. Higher velocity of the electrolyte is helpful for removing the products in machining and updating the electrolyte, which can improve machining localization and MRR. As a result, the specific air incidence angle leads to lower air void fraction, and a higher flow velocity is required in simulation.

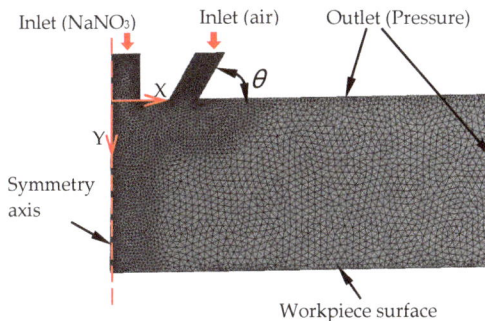

Figure 2. The structure model in the flow field.

3.2. Effects of Compressed Air Incidence Angle on Void Fraction (Air) in the Flow Field

Figure 3 shows the contours of void fraction (air) in the flow field by simulation at different incidence angles (θ) of compressed air. Blue and red colors represent the electrolyte and air, respectively. The cloud charts indicate that the electrolyte beam flowing into the machining gap was confined and becomes thinner with the decrease of the incidence angle (θ). Even, under the condition of $\theta = \pi/6$, the electrolyte beam disappears and becomes a mixture of electrolyte and air. That means that more air flows into the electrolyte beam and forms a mixed flow field to spray to the workpiece surface. In the machining process, more energy was expected to focus on the region where the material was removed. If there is more air mixing into the electrolyte, the void fraction of air would increase, which would lower the conductivity of the electrolyte, and so MRR decreases.

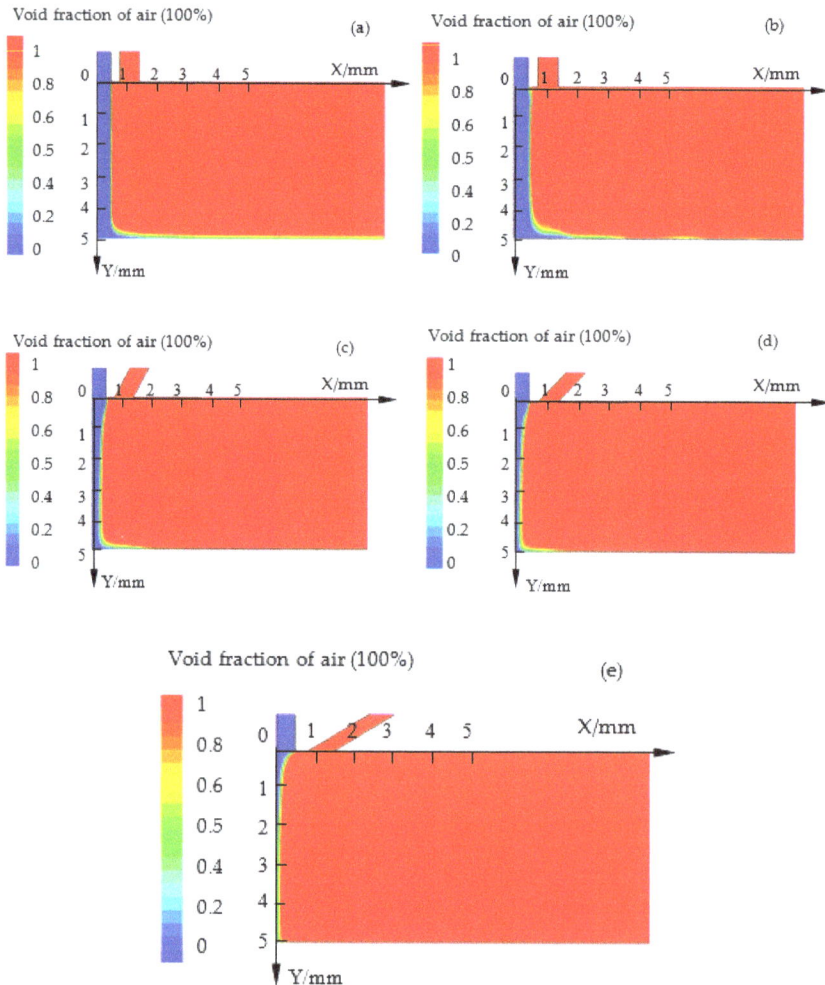

Figure 3. Contours of void fraction (air) in flow field by simulation at different incidence angles (θ): (a) No compressed air; (b) $\theta = \pi/2$; (c) $\theta = \pi/3$; (d) $\theta = \pi/4$; (e) $\theta = \pi/6$.

In order to obtain the value of the void fraction of air in the flow field, the data ($Y = 4, 0 < X < 2$) were extracted (see Figure 4) from the simulation model. The curve denotes that the void fraction of air increases, surrounding the symmetry Y axis, with the decrease of the incidence angle. When the incidence angle is larger than $\pi/4$, the focused effect is limited. However, under the condition of $\theta = \pi/6$, there is no pure electrolyte at the bottom of the cathode electrode. The electrolyte is a mixture of $NaNO_3$ and air, and this would lead to low MRR, which was not expected in the experiments. Therefore, based on the discussions above, the air incidence angle of $\theta = \pi/4$ was advised.

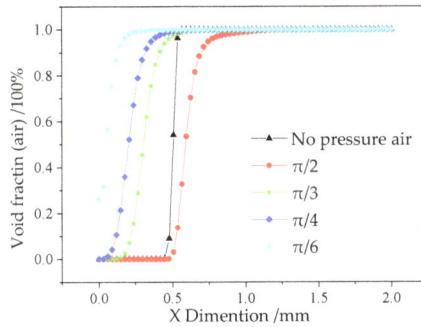

Figure 4. The extracted void fraction (air) (Y = 4, 0 < X < 2) at different incidence angles.

3.3. Effects of Compressed Air Incidence Angle on Velocity in the Flow Field

In EMM, the products (sludge, gas, and heat) in the machining gap should be removed at a certain electrolyte velocity. Timely renewal of the electrolyte can improve the machining speed and stability. It is also very important to remove the machining products and renew the electrolyte in the machining gap in AS-EMM. Figure 5 shows the contours of velocity in IEG by simulation at different incidence angles (θ). It can be seen that the $NaNO_3$ solution was shielded more with the decrease of the incidence angle. The velocity along the Y axis varies when the applied compressed air is different. Figure 6 depicts the graphs of velocity along the Y axis (X = 0) at different incidence angles. It shows that the velocity clearly increases with the decrease of incidence angle (θ) and it can be twice the velocity without compressed air (see Figure 5d,e, $\theta = \pi/4$ and $\pi/6$). The electrolyte would spray to the workpiece surface (Y = 5) at a higher speed and this would enhance the electrolyte renewal and discharge of the products. However, considering the void fraction of air, the smaller air incidence angle is a disadvantage to material removal, and $\theta = \pi/4$ is more appropriate.

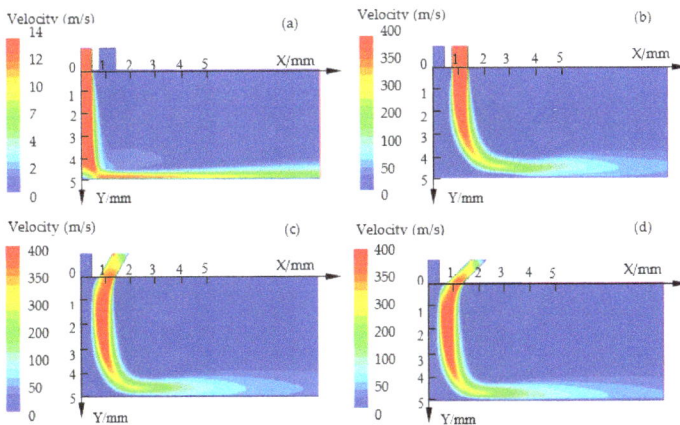

Figure 5. *Cont.*

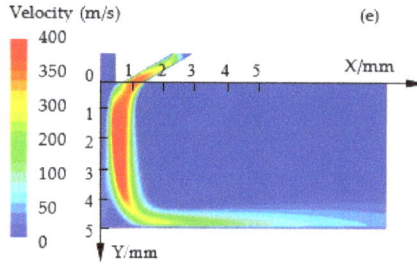

Figure 5. The electrolyte flow velocity in the interelectrode gap (IEG) in simulation at different incidence angles (θ): (**a**) No compressed air; (**b**) $\theta = \pi/2$; (**c**) $\theta = \pi/3$; (**d**) $\theta = \pi/4$; (**e**) $\theta = \pi/6$.

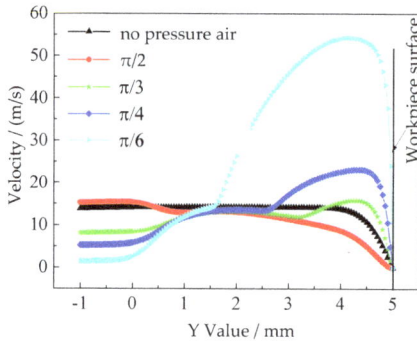

Figure 6. Graphs of velocity along the Y axis (X = 0) at different incidence angles (θ).

4. Experimental Design and Machining Results

The optimization of the processing parameters was carried out based on the advised nozzle inclination with the air incidence angle of $\pi/4$.

4.1. Design of Experiment

It was found that electrolyte velocity, gas velocity, and voltage had a great influence on machining results according to the preliminary experiments. In order to get the optimum processing parameter combination, an orthogonal test method was employed to design the experiments beforehand. The orthogonal test method is a kind of experimental design method to study multifactors and multilevels [21]. According to orthogonality, some representative points are selected from all experiments. It can greatly reduce the duration of the experiment and help get enough data rapidly. Therefore, a three-factor and four-level orthogonal experiment was design by Statistical Product and Service Solution (SPSS). The factors including electrolyte velocity, gas velocity and voltage, and their levels are shown in Table 1.

Table 1. Parameters and their levels.

Factors	Symbol	Unit	Levels			
			1	2	3	4
Electrolyte velocity	A	m/s	4	4.5	5	5.5
Gas velocity	B	m/s	0	80	120	160
Voltage	C	V	6	8	10	12

4.2. Experimental Details

The experimental condition is shown in Table 2. SS304 stainless steel (Mucheng, Foshan, China) and tungsten bar (Metalline, Luoyang, China) were selected as the workpiece material and tool material, respectively. Experiments were done to study the relationship between the machining parameters, including electrolyte velocity, gas velocity, and voltage, and the objectives, including the aspect ratio and the roughness (*Ra*) of the micro pit. Aspect ratio can be defined as $\lambda = H/D$ (*H* is the depth, and *D* is the diameter of micro pit). The profiles of pit were measured by a 3D profilometer (Olympus LEXT4500, Olympus, Tokyo, Japan). The roughness was measured by a surface roughness meter (SJ-411, Mitutoyo, Kawasaki, Japan) and its uncertainty is less than 5%.

Table 2. Parameters of experiment.

Parameters	Value
Machining gap	15 μm
Feed rate	60 μm·min^{-1}
Electrode diameter	50 μm
Electrolyte concentration	10% NaNO$_3$
Machining time	30 s
Duty ratio	50%
Frequency	100 kHz
Air incidence angle	π/4

4.3. Experimental Results

Scanning electron microscope (SEM) photos of the micro pits processed in AS-EMM, based on the conditions of Tables 1 and 2, are shown in Figure 7a–p. The measured data of aspect ratio and roughness for the micro pits from Figure 7a–p are listed in Table 3 (No. 1–16). It can be seen that the micro pit size and surface morphology processed by AS-EMM vary greatly under different experimental parameters. The profile of the pit is not obvious when the voltage and electrolyte velocity are too small due to less material removal (Figure 7a). However, excessive voltage and electrolyte velocity lead to large pit size, serious stray corrosion, and reduced roughness, as shown in Figure 7i. The average detected aspect ratio and the roughness obtained with 5 points at different parameters are listed in Table 3.

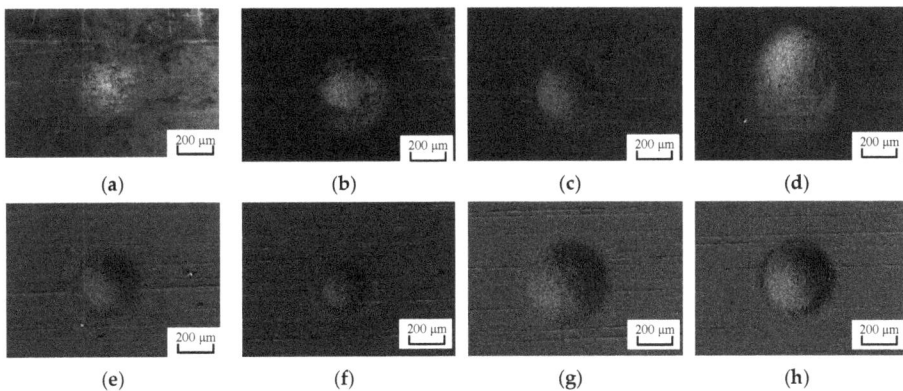

(a) (b) (c) (d)

(e) (f) (g) (h)

Figure 7. *Cont.*

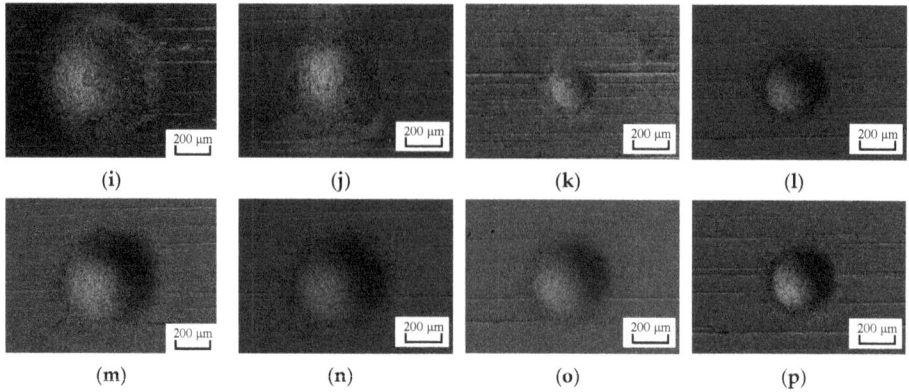

Figure 7. Morphology of machining results.

Table 3. The detected machining results.

Exp. No	A	B	C	λ	Ra/μm	Exp. No	A	B	C	λ	Ra/μm
1	1	1	1	0.064	0.116	9	3	1	3	0.176	0.113
2	1	2	2	0.099	0.121	10	3	2	4	0.190	0.167
3	1	3	3	0.184	0.128	11	3	3	1	0.132	0.139
4	1	4	4	0.174	0.152	12	3	4	2	0.219	0.081
5	2	1	2	0.140	0.139	13	4	1	4	0.215	0.187
6	2	2	1	0.126	0.097	14	4	2	3	0.196	0.159
7	2	3	4	0.188	0.138	15	4	3	2	0.204	0.072
8	2	4	3	0.201	0.136	16	4	4	1	0.181	0.043

5. Optimization of the Processing Parameters

The grey relational analysis is a method that can be applied to get the relationship between groups of data [22] and is fundamentally a simple and straightforward multicriteria decision-making technique [23]. To reduce errors caused by information asymmetry and better show influence of different parameters on machining results, grey relational analysis is introduced to determine the optimal parameters combination by analyzing the grey relational grade between parameters and responses.

5.1. Data Preprocessing

The grey relational analysis is based on the calculating the gray relational grade. It can be calculated according to the original processing results and then its relational coefficient can be computed. To simplify, it is necessary to make the original data dimensionless in units and orders of magnitude before calculating the grey relational grade.

The original sequence of 16 experimental results is set as $x_i(k)$, $i = 1, 2 \ldots \ldots , 16$; $k = 1, 2$.

Where i is the experimental sequence number, $k = 1$ represents the aspect ratio, and $k = 2$ represents the roughness.

As for the aspect ratio, in ECM, the larger the aspect ratio of pits is, the better localized the removal of the material is. Therefore, the pretreatment formula of aspect ratio is as follows:

$$x_i^*(k) = \frac{x_i(k) - \min x_i(k)}{\max x_i(k) - \min x_i(k)} \tag{1}$$

For surface roughness, the lower the machined surface roughness value is, the better the machining quality is. Therefore, the pretreatment formula of surface roughness is as follows:

$$x_i^*(k) = \frac{\max x_i(k) - x_i(k)}{\max x_i(k) - \min x_i(k)} \tag{2}$$

In Equations (1) and (2), $x_i(k)$, $x_i^*(k)$, $\max x_i(k)$, and $\min x_i(k)$ are the original sequence number, the sequence after data preprocessing, the maximum value, and minimum value in the original sequence, respectively. The dimensionless experimental results are shown in Table 4.

Table 4. Dimensionless experimental results.

Exp. No	Reference Sequence		Exp. No	Reference Sequence	
	λ	Ra		λ	Ra
1	0.000	0.493	9	0.724	0.514
2	0.225	0.458	10	0.811	0.139
3	0.772	0.410	11	0.438	0.333
4	0.711	0.243	12	1.000	0.736
5	0.491	0.333	13	0.977	0.000
6	0.397	0.625	14	0.853	0.194
7	0.797	0.340	15	0.901	0.799
8	0.886	0.354	16	0.752	1.000

5.2. Correlation Coefficient

According to Table 4, the grey relational coefficients of aspect ratio and surface roughness are respectively obtained. The calculation formula can be described as

$$\zeta(x_0(k), x_i^*(k)) = \frac{\min\limits_i \min\limits_k |x_0(k) - x_i^*(k)| + \rho \max\limits_i \max\limits_k |x_0(k) - x_i^*(k)|}{|x_0(k) - x_i^*(k)| + \rho \max\limits_i \max\limits_k |x_0(k) - x_i^*(k)|} \tag{3}$$

where ζ is the grey relational coefficient, ρ is distinguishing coefficient, which is usually 0.5, and $|x_0(k) - x_i^*(k)|$ is the absolute difference between the expected sequence (expected value is 1) and the sequence after data processing, where $x_0(k) = 1$ $(k = 1,2)$, $\min\limits_i \min\limits_k |x_0(k) - x_i^*(k)|$ is the minimum value of the absolute difference sequence, and $\max\limits_i \max\limits_k |x_0(k) - x_i^*(k)|$ is the maximum value of the absolute difference sequence.

The grey relational grade can be obtained by

$$\gamma_i = \frac{1}{N} \sum_{k=1}^{n} \zeta(x_0(k), x_i^*(k)) \tag{4}$$

where γ_i is the grey relational grade, and N is the number of performance characteristics.

The grey relational coefficient, grade of the aspect ratio, and roughness for each experiment were calculated by Equations (3) and (4) and listed in Table 5. Larger grey relational grade means a closer machining result to the desired value. From Table 5, the Exp. No. 12 has the highest grade 0.7887, which means the optimum parameter combination for the best structure morphology is A3B4C2.

Table 5. Grey relational coefficient and grey relational grade.

Exp. No	GRC		GRG	Rank	Exp. No	GRC		GRG	Rank
	λ	*Ra*				*λ*	*Ra*		
1	0.3333	0.4966	0.4149	16	9	0.6443	0.4286	0.5364	11
2	0.3920	0.4800	0.4360	15	10	0.7257	0.4513	0.5885	7
3	0.6864	0.4586	0.5725	8	11	0.4709	0.6145	0.5427	10
4	0.6340	0.3978	0.5159	12	12	1.0000	0.5774	0.7887	1
5	0.4954	0.4286	0.4620	14	13	0.9565	0.4977	0.7271	3
6	0.4534	0.5714	0.5124	13	14	0.7725	0.4777	0.6251	6
7	0.7115	0.4311	0.5713	9	15	0.8345	0.6341	0.7343	2
8	0.8145	0.4364	0.6254	5	16	0.6686	0.7294	0.6990	4

However, there are a total of 64 combinations in the three-factor and four-level experiments, among which there may be a parameter combination whose grey relational grade is higher than that of A3B4C2. Therefore, the influence of a single factor on the grey relational grade was calculated, which is shown in Figure 8. Based on Figure 8, the optimal combination of grey relational grade, which has the highest grade of each parameter, is A4B4C2, i.e., the electrolyte velocity is 5.5 m/s, the gas velocity is 160 m/s, and the voltage is 8 V.

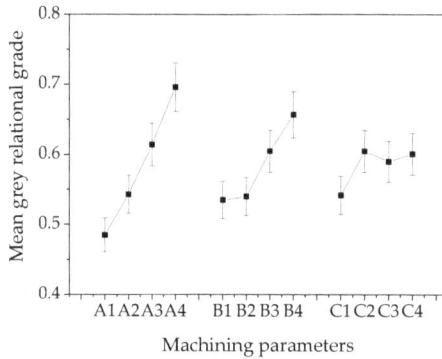

Figure 8. Mean grey relational grade of each parameter on machining results.

5.3. Experimental Verification

Verification experiments were carried out based on the optimum parameter combination of electrolyte velocity of 5.5 m/s, gas velocity of 160 m/s, and voltage of 8 V. The machining result is shown in Figure 9b. Figure 9 shows that the profile border of the machined pit by using the optimum parameters was sharper than that obtained by using the parameters in group 12. Its aspect ratio and roughness were 0.226 and 0.072, respectively, which is 3% higher and 11% lower than those of group 12. Verifying experiments denotes that the optimum parameters could be revised using the presented grey relational analysis method to realize a better machining morphology.

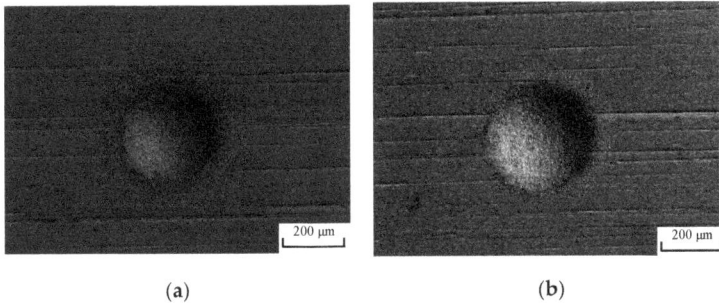

Figure 9. Micro pits machined using parameter combination of (a) group 12 and (b) the optimum group.

6. Conclusions

AS-EMM method was verified to be vital for machining microstructures on metal surface with high quality. In this paper, the nozzle inclination and machining parameters were optimized based on simulation and experiment analysis. Some conclusions are as follows:

1. Simulation results indicated that the void fraction (air) and velocity in the flow field increased with the decrease of nozzle inclination. According to the demonstrated model in this paper, the most appropriate nozzle inclination was $\theta = \pi/4$;

2. The optimal parameter combination for the multi-objective was A4B4C2; i.e., at 5.5 m/s electrolyte velocity, 160 m/s gas velocity, and 8 V voltage, the micro pit with better performance in aspect ratio and roughness could be processed at these conditions;

3. The proposed method makes a contribution to the improvement of quality of the micro pits in AS-EMM. It is also effective for optimization of the structure design and machining parameters in other machining methods.

Author Contributions: Conceptualization, M.W.; Methodology, M.W. and Y.S.; Software, Y.S.; Formal analysis, Y.S., K.H. and X.X.; Investigation, K.H.; Data curation, Y.S. and G.C.; Writing—original draft preparation, M.W. and Y.S.; Writing—review and editing, M.W. and Y.S.; Supervision, M.W.; Project administration, M.W.

Funding: This research was funded by (National Natural Science Foundation of China) grant numbers (51975532, 51475428) and (Natural Science Foundation of Zhejiang Province) grant numbers (LY19E050007).

Conflicts of Interest: The authors declare no conflict of interest.

References

1. Kim, Y.W.; Lee, J.M.; Lee, I.; Lee, S.H.; Ko, L.S. Skin friction reduction in tubes with hydrophobically structured surfaces. *Int. J. Precis. Eng. Manuf.* **2013**, *14*, 403–412. [CrossRef]

2. Huang, W.; Jiang, L.; Zhou, C.; Wang, X. The lubricant retaining effect of micro-dimples on the sliding surface of PDMS. *Tribol. Int.* **2012**, *52*, 87–93. [CrossRef]

3. Zhang, S.; Ochiai, M.; Sunami, Y.; Hashimoto, H. Influence of microstructures on aerodynamic characteristics for dragonfly wing in gliding flight. *J. Bionic Eng.* **2019**, *16*, 423–431. [CrossRef]

4. Bixler, G.D.; Bhushan, B. Bioinspired rice leaf and butterfly wing surface structures combining shark skin and lotus effects. *Soft Matter* **2012**, *8*, 11271–11284. [CrossRef]

5. Devaiah, M.; Santhosh Kumar, S.; Srihari, T.; Rajasekharan, T. SiCp/Al2O3 ceramic matrix composites prepared by directed oxidation of an aluminium alloy for wear resistance applications. *Trans. Indian Ceram. Soc.* **2012**, *71*, 151–158. [CrossRef]

6. Li, J.; Xiong, D.; Dai, J.; Huang, Z.; Tyagi, R. Effect of surface laser texture on friction properties of nickel-based composite. *Tribol. Int.* **2010**, *43*, 1193–1199. [CrossRef]

7. Xiao, N.; Khonsari, M.M. Thermal performance of mechanical seals with textured side-wall. *Tribol. Int.* **2012**, *45*, 1–7. [CrossRef]

8. Djamaï, A.; Brunetière, N.; Tournerie, B. Numerical modeling of thermohydrodynamic mechanical face seals. *Tribol. Trans.* **2010**, *53*, 414–425. [CrossRef]

9. Zhuang, J.; Hu, W.; Fan, Y.; Sun, J.; He, X.; Xu, H.; Wu, D. Fabrication and testing of metal/polymer microstructure heat exchangers based on micro embossed molding method. *Microsyst. Technol.* **2019**, *25*, 381–388. [CrossRef]

10. Jain, V.K.; Sidpara, A.; Balasubramaniam, R.; Lodha, G.S.; Dhamgaye, V.P.; Shukla, R. Micromanufacturing: A review—Part I. *Proc. Inst. Mech. Eng. B-J. Eng.* **2014**, *228*, 973–994. [CrossRef]

11. Liu, Z.; Nouraei, H.; Spelt, J.K.; Papini, M. Electrochemical slurry jet micro-machining of tungsten carbide with a sodium chloride solution. *Precis. Eng.* **2015**, *40*, 189–198. [CrossRef]

12. Kunieda, M.; Mizugai, K.; Watanabe, S.; Shibuya, N.; Iwamoto, N. Electrochemical micromachining using flat electrolyte jet. *CIRP Ann. Manuf. Technol.* **2011**, *60*, 251–254. [CrossRef]

13. Schubert, A.; Hackert-Oschätzchen, M.; Martin, A.; Winkler, S.; Kuhn, D.; Meichsner, G.; Zeidler, H.; Edelmann, J. Generation of complex surfaces by superimposed multi-dimensional motion in electrochemical machining. *Procedia CIRP* **2016**, *42*, 384–389. [CrossRef]

14. Clare, A.T.; Speidel, A.; Bisterov, I.; Jackson, A. Precision enhanced electrochemical jet processing. *CIRP Ann.* **2018**, *67*, 205–208. [CrossRef]

15. Hackert-Oschätzchen, M.; Paul, R.; Martin, A.; Meichsner, G.; Lehnert, N.; Schubert, A. Study on the dynamic generation of the jet shape in jet electrochemical machining. *J. Mater. Process. Technol.* **2015**, *223*, 240–251. [CrossRef]

16. Zhao, Y.H.; Masanori, K. Investigation on electrolyte jet machining of three-dimensional freeform surfaces. *Precis. Eng.* **2019**, *60*, 42–53. [CrossRef]

17. Chen, X.L.; Dong, B.Y.; Zhang, C.Y.; Wu, M.; Guo, Z.N. Jet electrochemical machining of micro dimples with conductive mask. *J. Mater. Process. Technol.* **2018**, *257*, 101–111. [CrossRef]

18. Speidel, A.; Mitchell-Smith, J.; Walsh, D.A.; Hirsch, M.; Clare, A.T. Electrolyte jet machining of titanium alloys using novel electrolyte solutions. *Procedia CIRP* **2016**, *42*, 367–372. [CrossRef]

19. Wang, M.H.; Bao, Z.Y.; Qiu, G.Z.; Xu, X.F. Fabrication of micro-dimple arrays by AS-EMM and EMM. *Int. J. Adv. Manuf. Technol.* **2017**, *93*, 787–797. [CrossRef]

20. Wang, M.H.; Tong, W.J.; Qiu, G.Z.; Xu, X.F.; Speidle, A.; Mitchell-Smith, J. Multiphysics study in air-shielding electrochemical micromachining. *J. Manuf. Process.* **2019**, *43*, 124–135. [CrossRef]

21. Xia, S.; Lin, R.; Cui, X.; Shan, J. The application of orthogonal test method in the parameters optimization of PEMFC under steady working condition. *Int. J. Hydrogen Energy* **2016**, *41*, 11380–11390. [CrossRef]

22. Malik, A.; Manna, A. Multi-response optimization of laser-assisted jet electrochemical machining parameters based on gray relational analysis. *J. Braz. Soc. Mech. Sci. Eng.* **2018**, *40*, 148. [CrossRef]

23. Singh, T.; Patnaik, A.; Chauhan, R. Optimization of tribological properties of cement kiln dust-filled brake pad using grey relation analysis. *Mater. Des.* **2016**, *89*, 1335–1342. [CrossRef]

micromachines

MDPI

Article

Effects of Machining Errors on Optical Performance of Optical Aspheric Components in Ultra-Precision Diamond Turning

Yingchun Li [1], Yaoyao Zhang [1], Jieqiong Lin [1,*], Allen Yi [2] and Xiaoqin Zhou [3]

[1] School of Mechatronic Engineering, Changchun University of Technology, Changchun 130012, China; chun04230525@126.com (Y.L.); 2201801055@stu.ccut.edu.cn (Y.Z.)

[2] Department of Industrial, Welding and systems Engineering, Ohio State University, Columbus, OH 43210, USA; yi.71@osu.edu

[3] School of Mechanical and Aerospace Engineering, Jilin University, Changchun 130022, China; xqzhou@jlu.edu.cn

* Correspondence: linjieqiong@ccut.edu.cn; Tel.: +86-137-5606-7918

Received: 6 February 2020; Accepted: 19 March 2020; Published: 23 March 2020

Abstract: Optical aspheric components are inevitably affected by various disturbances during their precision machining, which reduces the actual machining accuracy and affects the optical performance of components. In this paper, based on the theory of multi-body system, we established a machining error model for optical aspheric surface machined by fast tool servo turning and analyzed the effect of the geometric errors on the machining accuracy of optical aspheric surface. We used the method of ray tracing to analyze the effect of the surface form distortion caused by the machining error on the optical performance, and identified the main machining errors according to the optical performance. Finally, the aspheric surface was successfully applied to the design of optical lens components for an aerial camera. Our research has a certain guiding significance for the identification and compensation of machining errors of optical components.

Keywords: optical aspheric surface; fast tool serve (FTS); machining error; optical performance

1. Introduction

As a new type of optical surface, an optical aspheric surface has obvious advantages, including correcting aberration, reducing the size and weight of the system, expanding the field of view, compared with a traditional regular surface, and has become a core component of modern optical systems [1–3]. In order to meet the actual optical performance of the components, it is necessary to rely on high-precision machining. However, the processing of components will inevitably be affected by various factors, such as geometric error, tool error, and thermal error [4,5]. All error components are reflected on the surface of the workpiece through the kinematic chain of the machine tool, which causes the form distortion of actual machined surface and affects optical performance of the components [6]. Therefore, in order to improve the machining accuracy of aspheric components, it is necessary to perform an error analysis on the machining process.

Usually, the geometric error is a basic factor influencing the machining accuracy [7]. Leete [8], based on trigonometric relationship, established a geometric error model for a three-axis computer numerical control (CNC) machine tools. Based on the assumption of rigid body motion and small angle error, Ferreira [9] proposed an analytical model for the prediction of geometric error of a three-axis machine tool. Okafor [10] used homogeneous transformation matrix (HTM) to model and analyze the geometric error and thermal error of the vertical three-axis machining center. Lamikiz et al. [11] used Denavit–Hartenberg (D–H) convention to model the geometric error of the machining center. Based on the theory of multi-body system (MBS), Kong et al. [12] established a volumetric error model for ultra-precision machine tools. In addition, some scholars use neural network theory and stream of variation theory to analyze machine tool errors [13,14].

Due to the existence of geometric error of machine tool, the form distortion will occur on the surface of the workpiece being machined. The sensitivity of the geometric errors' effect on form error is different for machining surface, so the compensation of the errors with high sensitivity is more effective [15]. Li et al. [16] used the matrix differential method to study the effect of geometric error on machining accuracy. Cheng et al. [17] identified main geometric errors for a multi-axis machine tool based on the MBS theory. Note that in previous studies, the evaluation of machining quality of components only relies on geometric form. However, as an applied device, the processing of optical aspheric components should meet the needs of optical applications [18].

The main evaluation parameters of optical performance for optical surfaces include wavefront aberration [19,20], modulation transfer function (MTF) [21], point spread function (PSF) [22] and power spectral density (PSD) [23,24], etc. As a comprehensive index, wavefront aberration can transform with other evaluation parameters [18], so the evaluation parameter based on wavefront aberration was applied in this paper to study the influence of machining errors.

In this paper, based on the mothed of creating optical aspheric components by FTS turning, we studied the geometric error modeling of the machine tool, the modeling of three-dimensional topography for machined surface, and the evaluation modeling of optical performance. The aim of this paper was to establish the relationship between the geometric error of machine tool and the form error of machined surface and the optical performance of aspheric surface, and to realize the identification of main machining errors based on optical performance, so as to guide the machining for optical aspheric components with specific optical performance and promote the wide application of aspheric surface.

2. Volumetric Error Modeling

It is important for the analysis, separation and compensation of the machining errors to establish the volumetric error model of the machine tool [25,26]. The schematic diagram of FTS turning system is shown in Figure 1a. In the process of machining for aspheric surfaces, the tool reciprocates linearly in the axial direction under the driving of the FTS, which can be viewed as a prismatic pair along Z-direction. Therefore, the system includes three translational axes and one rotational axis, i.e., X axis, Z axis, FTS axis and C axis, respectively.

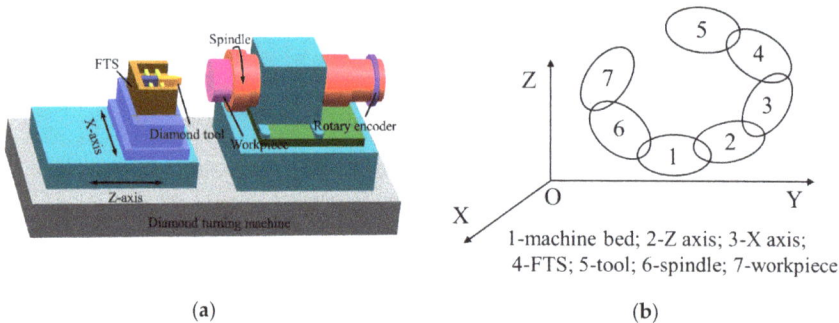

1-machine bed; 2-Z axis; 3-X axis;
4-FTS; 5-tool; 6-spindle; 7-workpiece

(a)   (b)

Figure 1. (**a**) Schematic diagram and (**b**) Kinematic chain diagram of the fast tool serve (FTS) turning system.

The number of error components depending on the number of axes of a particular machine. Since each axis of the machine system has six degrees of freedom, it has six error components. From the geometric errors analysis, it can be seen that the FTS turning system has 24 error components provided by X-axis, Z-axis, C-axis and FTS-axis and 6 squareness errors. However, in order to simplify the error model, only the squareness error between X-axis and Z-axis were considered among the 6 squareness errors. Table 1 shows the 26 geometric errors of the FTS turning system.

Table 1. Geometric error components of the fast tool serve (FTS) turning system.

Axis	Error Terms
Axis X	$\delta_{xx}, \delta_{xy}, \delta_{xz}, \varepsilon_{xx}, \varepsilon_{xy}, \varepsilon_{xz}$
Axis Z	$\delta_{zx}, \delta_{zy}, \delta_{zz}, \varepsilon_{zx}, \varepsilon_{zy}, \varepsilon_{zz}$
Spindle (Axis C)	$\delta_{cx}, \delta_{cy}, \delta_{cz}, \varepsilon_{cx}, \varepsilon_{cy}, \varepsilon_{cz}$
Axis FTS	$\delta_{fx}, \delta_{fy}, \delta_{fz}, \varepsilon_{fx}, \varepsilon_{fy}, \varepsilon_{fz}$
Squareness error	η_{zx}

δ_{mn}: Displacement errors; ε_{mn}: Angular error, where the first subscript refers to the motion axis, the second subscript refers to the error direction or the rotation axis of angular error, η_{zx}: the squareness error between axis X and axis Z.

2.1. Transformation Matrix between Adjacent Bodies

The relative location and attitude between two bodies can be obtained through HTM based on the MBS theory [27]. The topology of the FTS turning system is shown in Figure 1b, which includes a tool branch form the machine bed to the cutting tool and a workpiece branch from the machine bed to the workpiece.

When machining aspheric surfaces with FTS turning, the Z-axis carriage of machine tool is only used to calibrate the initial position of the X-axis carriage and does not participate in the machining motion. Therefore, the connection between the Z-axis carriage and the machine bed be regarded as static, that the HTM $T_{12} = I_{4\times4}$. Moreover, there is no relative movement between the tool and FTS, that is $T_4^5 = I_{4\times4}$.

The misalignments between the X-axis carriage and the Z-axis carriage cause a small offset called the squareness error η_{zx}. Therefore, the HTM T_2^3 between the X-axis carriage and the Z-axis carriage should take into account the effect of the squareness error η_{zx}.

$$T_2^3 = \begin{bmatrix} 1 & -\varepsilon_{xz} & \varepsilon_{xy} & R-x+\delta_{xx} \\ \varepsilon_{xz} & 1 & -\varepsilon_{xx} & \delta_{xy} \\ -\varepsilon_{xy} & \varepsilon_{xx} & 1 & \delta_{xz}-x\eta_{zx} \\ 0 & 0 & 0 & 1 \end{bmatrix} \tag{1}$$

where R represents the radius of the workpiece being machined and x represents the amount of movement of the X-axis carriage during processing.

The FTS device is mounted on the X-axis carriage, and the tool holder coupled to it moves in the Z-direction respect to the X-axis carriage. So the HTM T_3^4 from the X-axis carriage to the FTS can be formulated as follows after considering the error components listed in Table 1.

$$T_3^4 = \begin{bmatrix} 1 & -\varepsilon_{fz} & \varepsilon_{fy} & \delta_{fx} \\ \varepsilon_{fz} & 1 & -\varepsilon_{fx} & \delta_{fy} \\ -\varepsilon_{fy} & \varepsilon_{fx} & 1 & S-z_f+\delta_{fz} \\ 0 & 0 & 0 & 1 \end{bmatrix} \tag{2}$$

where S represents the surface Sag value of the optical aspheric component being machined; z_f represents the amount of movement of the tool holder in the Z direction during the processing.

In practice, the actual rotation center of the spindle will shift from the nominal rotation center. When the spindle rotates θ angle relative to the bed, the HTM T_1^6 between them can be represented as Equation (3) with the assumption of small angle approximation.

$$T_1^6 = \begin{bmatrix} cos\theta - \varepsilon_{cz}sin\theta & -sin\theta - \varepsilon_{cz}cos\theta & \varepsilon_{cy}cos\theta + \varepsilon_{cx}sin\theta & \delta_{cx}cos\theta - \delta_{cy}sin\theta \\ sin\theta + \varepsilon_{cz}cos\theta & cos\theta - \varepsilon_{cz}sin\theta & \varepsilon_{cy}sin\theta - \varepsilon_{cx}cos\theta & \delta_{cy}cos\theta + \delta_{cx}sin\theta \\ -\varepsilon_{cy} & \varepsilon_{cx} & 1 & \delta_{cz} \\ 0 & 0 & 0 & 1 \end{bmatrix} \tag{3}$$

Since the workpiece is fixed on the spindle, the HTM T_6^7 from the workpiece to the spindle is also a unitary matrix, that is, $T_6^7 = I_{4\times4}$.

2.2. Integrated Volumetric Errors Model

The process of developing the integrated errors model mainly aims to obtain the relative displacement error between the cutting tool and the workpiece in the turning process [28]. Assume that the cutting tool tip position in tool coordinate system (TCS) is as follows.

$$P_t = [x_t, \ y_t, \ z_t, \ 1]^T \tag{4}$$

Then, the cutting tool position in the workpiece coordinate system (WCS) can be formulated as

$$P_w = \left(_1^7T\right)^{-1} {}_1^5T P_t \tag{5}$$

where

$${}_1^7T = {}_1^6T {}_6^7T \tag{6}$$

$${}_1^5T = {}_1^2T {}_2^3T {}_3^4T {}_4^5T \tag{7}$$

Under ideal conditions, all the errors in Table 1 are equal to zero. Combining the Equations (6) and (7), the ideal form-shaping function (the ideal cutting tool position in the WCS) can be expressed as Equation (8):

$$P_{w0} = \left(_1^7T_0\right)^{-1} \left(_1^5T_0\right) P_t$$

$$= \begin{bmatrix} cos\theta & -sin\theta & 0 & 0 \\ sin\theta & cos\theta & 0 & 0 \\ 0 & 0 & 1 & 0 \\ 0 & 0 & 0 & 1 \end{bmatrix}^{-1} \begin{bmatrix} 1 & 0 & 0 & R-x \\ 0 & 1 & 0 & 0 \\ 0 & 0 & 1 & 0 \\ 0 & 0 & 0 & 1 \end{bmatrix} \begin{bmatrix} 1 & 0 & 0 & 0 \\ 0 & 1 & 0 & 0 \\ 0 & 0 & 1 & S-Z_f \\ 0 & 0 & 0 & 1 \end{bmatrix} \begin{bmatrix} x_t \\ y_t \\ z_t \\ 1 \end{bmatrix} \tag{8}$$

In the actual machining process, the cutting tool point is the combination of the ideal cutting tool point and the error motion [29]. Therefore, considering the geometric error terms of machine tool in Table 1, the actual shape-forming function (the actual cutting tool position in the WCS) is formulated as Equation (9):

$$
P_w = \left(^7_1T\right)^{-1}\left(^5_1T\right)P_t =
\left(\left(\begin{bmatrix} \cos\theta & -\sin\theta & 0 & 0 \\ \sin\theta & \cos\theta & 0 & 0 \\ 0 & 0 & 1 & 0 \\ 0 & 0 & 0 & 1 \end{bmatrix}\begin{bmatrix} 1 & -\varepsilon_{cz} & \varepsilon_{cy} & \delta_{cx} \\ \varepsilon_{cz} & 1 & -\varepsilon_{cx} & \delta_{cy} \\ -\varepsilon_{cy} & \varepsilon_{cx} & 1 & \delta_{cz} \\ 0 & 0 & 0 & 1 \end{bmatrix}\right)^{-1}\right.
$$

$$
= \left(\begin{bmatrix} 1 & 0 & 0 & R-x \\ 0 & 1 & 0 & 0 \\ 0 & 0 & 1 & 0 \\ 0 & 0 & 0 & 1 \end{bmatrix}\begin{bmatrix} 1 & -\varepsilon_{xz} & \varepsilon_{xy} & \delta_{xx} \\ \varepsilon_{xz} & 1 & -\varepsilon_{xx} & \delta_{xy} \\ -\varepsilon_{xy} & \varepsilon_{xx} & 1 & \delta_{xz}+x\eta_{zx} \\ 0 & 0 & 0 & 1 \end{bmatrix}\right)
$$

$$
\left.\begin{bmatrix} 1 & 0 & 0 & 0 \\ 0 & 1 & 0 & 0 \\ 0 & 0 & 1 & S-Z_f \\ 0 & 0 & 0 & 1 \end{bmatrix}\begin{bmatrix} 1 & -\varepsilon_{fz} & \varepsilon_{fy} & \delta_{fx} \\ \varepsilon_{fz} & 1 & -\varepsilon_{fx} & \delta_{fy} \\ -\varepsilon_{fy} & \varepsilon_{fx} & 1 & \delta_{fz} \\ 0 & 0 & 0 & 1 \end{bmatrix}\right)\begin{pmatrix} x_t \\ y_t \\ z_t \\ 1 \end{pmatrix}
\tag{9}
$$

Combining Equations (8) and (9), the volumetric error E of FTS turning system can be obtained.

$$
E = \left[E_x, E_y, E_z, 1\right]^T = P_w - P_{w0} \tag{10}
$$

Finally, the position deviation of the cutting tool in the WCS can be expressed as follows (ignore the higher-order terms):

$$
\begin{aligned}
E_x =\ & \left(\delta_{fx}+\delta_{xx}\right)\cos\theta + \left(\delta_{fy}+\delta_{xy}\right)\sin\theta - \varepsilon_{xx}\left(S-z_f\right)\sin\theta \\
& +\varepsilon_{xy}\left(S-z_f\right)\cos\theta + \varepsilon_{cz}(x-R)\sin\theta - \delta_{cx} - \varepsilon_{cy}\left(S-z_f\right) \\
E_y =\ & \left(-\delta_{fx}-\delta_{xx}\right)\sin\theta + \left(\delta_{fy}+\delta_{xy}\right)\cos\theta - \varepsilon_{xx}\left(S-z_f\right)\cos\theta \\
& -\varepsilon_{xy}\left(S-z_f\right)\sin\theta + \varepsilon_{cz}(x-R)\cos\theta - \delta_{cy} + \varepsilon_{cx}\left(S-z_f\right) \\
E_z =\ & \left(\delta_{xz}+\delta_{fz}-\delta_{cz}\right) + \varepsilon_{cx}(R-x)\sin\theta + \varepsilon_{cy}(R-x)\cos\theta - H_{zx}x
\end{aligned}
\tag{11}
$$

3. The Influence of Machining Errors

According to the effect of the error components on the coordinate distortions [29], the error components in Equation (11) are simplified to 11 errors as shown in Table 2. The influence of 11 simplified errors on the coordinate distortions is different.

Table 2. The simplified model of the machining errors.

Error Terms	Coordinate Distortion in the X Direction E_x	Coordinate Distortion in the Y Direction E_y	Coordinate Distortion in the Z Direction E_z
$\delta_x = \delta_{fx}+\delta_{xx}$	$\delta_x\cos\theta$	$-\delta_x\sin\theta$	0
$\delta_y = \delta_{fy}+\delta_{xy}$	$\delta_y\sin\theta$	$\delta_y\cos\theta$	0
$\delta_z = \delta_{xz}+\delta_{fz}-\delta_{cz}$	0	0	δ_z
ε_{xx}	$-\varepsilon_x\left(S-z_f\right)\sin\theta$	$-\varepsilon_x\left(S-z_f\right)\cos\theta$	0
ε_{xy}	$\varepsilon_y\left(S-z_f\right)\cos\theta$	$-\varepsilon_y\left(S-z_f\right)\sin\theta$	0
ε_{cz}	$\varepsilon_z(x-R)\sin\theta$	$\varepsilon_z(x-R)\cos\theta$	0
δ_{cx}	$-\delta_{cx}$	0	0
δ_{cy}	0	$-\delta_{cy}$	0
ε_{cx}	0	$\varepsilon_{cx}\left(S-z_f\right)$	$\varepsilon_{cx}(R-x)\sin\theta$
ε_{cy}	$-\varepsilon_{cy}\left(S-z_f\right)$	0	$\varepsilon_{cy}(R-x)\cos\theta$
H_{zx}	0	0	$-H_{zx}x$

These main error components should be considered in the machining and compensation process. Therefore, 12 sets of simulation cases are constructed to study the influence of geometric error components in the process of machining aspheric surface with FTS turning. Due to the distribution trend of the coordinate distortions will not change with the change of error values, the position errors of machine tool are generally less than 0.001mm, and the angle errors are generally less than 0.001°, so all error values are set to 0.001mm (0.001°) for simulation. The simulated plans are listed in Table 3.

Table 3. Simulation plans and error values.

Case No.	δ_x (mm)	δ_y (mm)	δ_z (mm)	ε_{xx} (mm)	ε_{xy} (mm)	ε_{cz} (mm)	δ_{cx} (mm)	δ_{cy} (mm)	ε_{cx} (mm)	ε_{cy} (mm)	H_{zx} (mm)
A	0.001	0	0	0	0	0	0	0	0	0	0
B	0	0.001	0	0	0	0	0	0	0	0	0
C	0	0	0.001	0	0	0	0	0	0	0	0
D	0	0	0	0.001	0	0	0	0	0	0	0
E	0	0	0	0	0.001	0	0	0	0	0	0
F	0	0	0	0	0	0.001	0	0	0	0	0
G	0	0	0	0	0	0	0.001	0	0	0	0
H	0	0	0	0	0	0	0	0.001	0	0	0
I	0	0	0	0	0	0	0	0	0.001	0	0
J	0	0	0	0	0	0	0	0	0	0.001	0
K	0	0	0	0	0	0	0	0	0	0	0.001
L	0.001	0.001	0.001	0.001	0.001	0.001	0.001	0.001	0.001	0.001	0.001

3.1. Effect on Form Distortion

Due to the existence of the coordinate distortions in the X, Y and Z directions caused by the machining errors, there is a form deviation between the actual (considering the error terms in Table 1) surface and ideal (each error terms in Table 1 is 0) surface. According to the plans in Table 3, simulation is performed using a toric surface to study the effect of machining error components on the form distortion of optical aspheric surface. The toric surface is expressed as Equation (12).

$$z = R_b - \sqrt{\left(R_b - R_d + \sqrt{R_d^2 - \rho^2 sin^2\theta}\right)^2 - \rho^2 cos^2\theta} \tag{12}$$

where $R_b = 265\ mm$ and $R_d = 132.5\ mm$ are set in this article.

In the simulation, the tool arc radius is set as r = 0.5 mm, the spindle speed is N = 500 r/mm and the feed rate is set as $a_f = 0.02$ mm/r. Under ideal conditions, the turning surface topography is shown in Figure 2a. The rotary asymmetry and bisymmetry characteristics of the toric surface can be seen clearly, which is the desired surface under the Cartesian coordinate system.

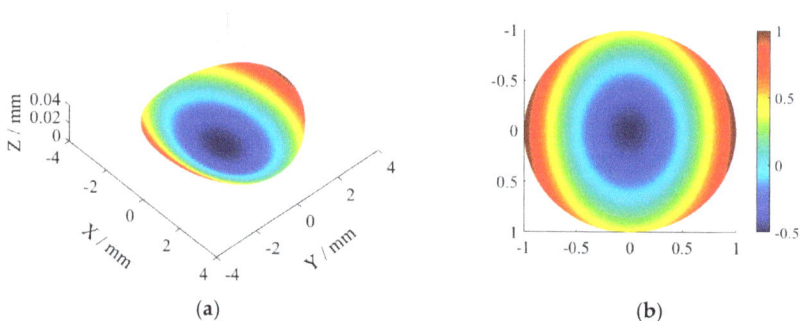

(a)

(b)

Figure 2. Under ideal conditions, toric surface's (a) Three-dimensional topography map of the machined surface and (b) Wavefront map.

Under the actual machining situation, the existence of geometric errors of machining system will affect the shape accuracy of the machined surface for components, and the influence of different geometric errors is also different. The main error components should be given priority in the processing and compensation of components. Figure 2a shows the contribution of the machining error terms to the form distortion by calculating the root mean square deviation (S_q) values of the form errors. Among them, the form distortion data is obtained from the surface topography by wavelet analysis, which only contains the low-frequency components.

Principal component analysis can be used to find the main error terms [28]. As can be from Figure 3a, the contribution of error δ_z is significantly larger than that of other errors in the figure, which is the main machining error term affecting the form accuracy.

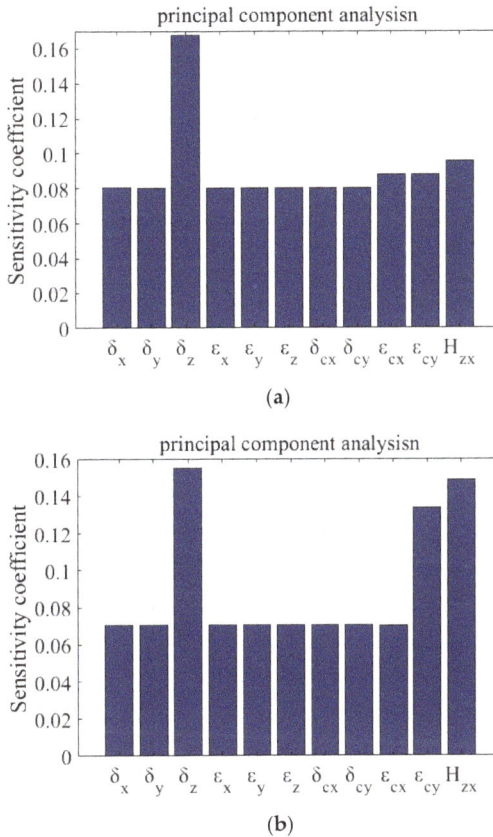

(a)

(b)

Figure 3. The contribution of different error components: (**a**) To form distortion; (**b**) To wavefront increment.

3.2. Effect on Optical Performance

The form distortion of optical aspheric components will affect their own optical performance. Wavefront aberration is the optical path difference between the actual wavefront and the ideal wavefront, form which one can easily derive MTF, PSF, and other optical parameters [18,30]. In this paper, the wavefront aberration was obtained by using the Zernike polynomials to fit the discrete data with the same phase after the light is refracted (or reflected).

Figure 2b shows the wavefront map corresponding to the machined toric surface under the ideal conditions. Under the actual machining situation, for aspheric surface, the form distortions caused by different machining error components are different, and different form distortions will also cause different aberrations. It has been proven that although the form distortions of the two surfaces had similar peak to valley (PV) values, their final optical performances were obviously different due to the different distribution of the errors [26]. Therefore, the principal component analysis of machining errors based on optical performance is performed in this paper.

The increment of wavefront aberration is the deviation of the wavefront under the condition with error disturbances from that the wavefront under the ideal machining condition. As shown in Figure 3b, the contribution of the geometric error components to the wavefront increment by calculating the S_q values of the wavefront aberration increment. It can be seen from the figure that the contributions of error H_{zx} and error ε_{cy} to wavefront increment are close to that of error δ_z , which are all the main machining errors that affect the optical performance of the components. In combination with Figure 3a, it can be concluded that the form distortions caused by the error H_{zx} and the error ε_{cy} will greatly reduce the optical performance of the components compared with the form distortions caused by other errors. Error H_{zx} , ε_{cy} , and δ_z are all the main machining error terms of toric surface created with FTS turning.

3.3. Details of the Effect on Optical Performace

In this paper, since the wavefront was obtained by fitting using Zernike polynomials and Zernike polynomials correspond to the primary aberrations, wavefront aberration is a comprehensive reflection of multiple geometric aberrations. The size of Zernike fitting coefficients symbolizes the size of different geometric aberrations. From Figure 3, it can be seen that the error components δ_z , H_{zx} and ε_{cy} are the main machining error terms in the process of creating toric surface with FTS turning. Therefore, the first nine items of the fitting coefficients of the wave surface under ideal condition and the wave surface under cases C, J and K are listed in the Table 4 to further analyze the effect of machining errors on optical performance.

Table 4. Zernike fitting coefficients of the wavefront of machined toric surface under different conditions.

Coefficient No. q_i	Ideal Coefficients	Actual Coefficients under the Influence of Different Errors		
		δ_z	ε_{cy}	H_{zx}
q_1	0.017775607	0.177403762	0.17765288	0.180462849
q_2	2.60×10^{-18}	3.10×10^{-7}	3.32×10^{-7}	2.74×10^{-7}
q_3	-6.59×10^{-5}	-0.002374094	-0.00244473	-0.002048713
q_4	4.22×10^{-18}	2.86×10^{-7}	3.06×10^{-7}	2.53×10^{-7}
q_5	0.311282774	0.34964832	0.35004824	0.3513772
q_6	-0.148267872	-0.128465135	-0.128461443	-0.128424573
q_7	4.33×10^{-18}	-1.35×10^{-7}	-1.38×10^{-7}	-1.21×10^{-7}
q_8	9.10×10^{-18}	2.51×10^{-7}	2.41×10^{-7}	2.25×10^{-7}
q_9	-1.28×10^{-5}	-0.003751194	-0.003833412	-0.003912768

It can be seen from Table 4 that due to the influence of machining errors, the fitting coefficients of the wavefront of the actual toric surface is larger than the ideal coefficients, that is, in the actual machining process of the optical aspheric components, the existence of the geometric errors will increase the wavefront aberration and the primary aberrations of the components and reduce the optical performance of the optical components. In addition, by comparing the values of fitting coefficients of Zernike polynomials under different conditions, it can be concluded that the variation of coefficients q_5 and q_6 is larger than that of other coefficients, which correspond to the focus and Y-astigmatism, that is, the machining errors has the greatest impact on these two aberrations.

In order to further clarify the details of the influence on geometric aberration, a principal component analysis was carried out to identify the main machining errors that affect the focus and the Y-astigmatism of the optical aspheric surface. As shown in Figure 4, the maps of focus and Y-astigmatism under the influence of three main machining errors of δ_z , H_{zx} and ε_{cy} , and Figure 5 is the analysis of main machining errors affecting the two aberrations.

Focus   Y-astigmatism

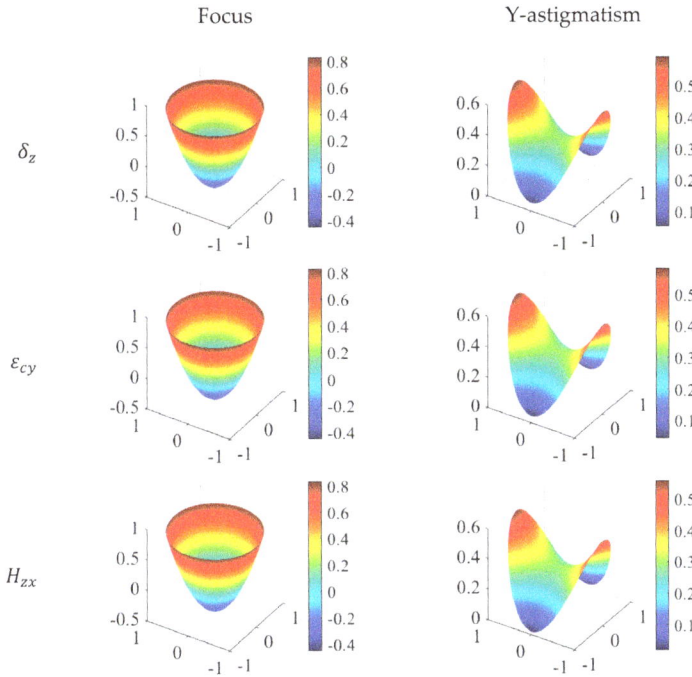

Figure 4. The focus map and the Y-astigmatism map under the influence of errors δ_z, ε_{cy} and H_{zx}.

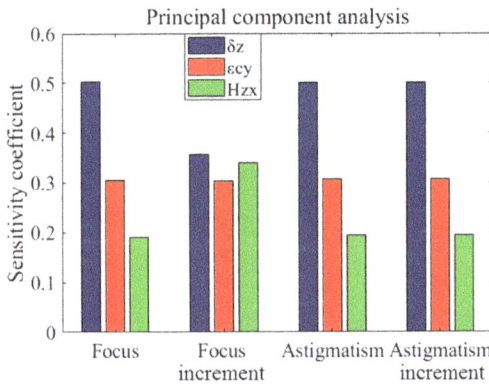

Figure 5. The contribution of three main error components δ_z, ε_{cy} and H_{zx}.

The following conclusions can be drawn from Figures 4 and 5.

(1) The machining error δ_z is not only the main contributor to the wavefront aberration, but also the main contributor to focus and Y-astigmatism;

(2) According to the contribution to focus, the three machining errors can be sorted, $\delta_z > \varepsilon_{cy} > H_{zx}$, but they have the same impact on the focus increment.

(3) The contributions of the three main machining errors to the Y-astigmatism and the increment of Y-astigmatism all can be sorted, $\delta_z > \varepsilon_{cy} > H_{zx}$.

(4) In the process of machining aspheric surface with FTS turning, we should first control and compensate the geometric error component δ_z (δ_{xz}, Z-direction displacement error of X-axis; δ_{fz},

Z-direction displacement error of FTS-axis; δ_{cz} , Z-direction displacement error of spindle) of the lathe, and then ε_{cy} (Y-direction angular error of spindle) and H_{zx} (the squareness error between the X and Z axis).

4. Compensation for Main Machining Errors

The form quality and optical performance of the machined surface mainly depend on the main machining errors. In Section 3, the main error components in the actual machining process of optical aspheric surface with FTS turning are δ_z, H_{zx} and ε_{cy} . Therefore, in order to improve the quality of machined surface, the main machining error components are compensated in this section.

In Equation (11), the position deviation of the cutting tool under the influence of the errors is shown. It can be seen that the three main geometric error components mainly affect the Z coordinate, and distortion values is also obtained. Error compensation requires that a value equal to the distortion be superimposed in the direction opposite to the distortion direction, i.e., E_z . The identified machining errors were compensated by modifying the tool path to improve the form equality and the optical performance in this section. As shown in Figure 6a, the form error before compensation for the machined toric surface, with a S_q value of 9.25 µm. The form error after compensation, with a S_q value of 2.26 µm, is shown in Figure 6b. Compared to the form accuracy before compensation, the accuracy is improved by 82.29%.

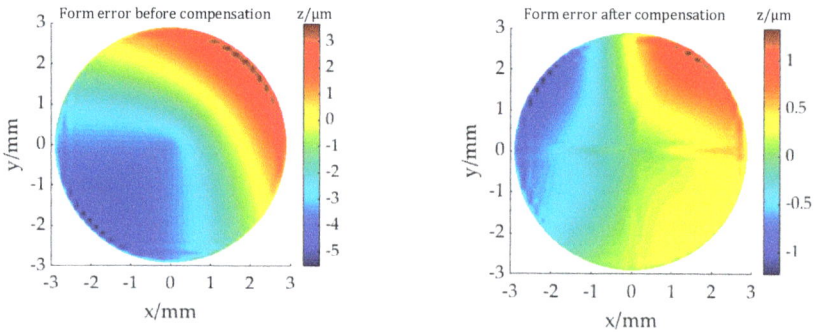

Figure 6. Form error of the machined toric surface: (**a**) before compensation; (**b**) after compensation.

Zernike polynomial coefficient corresponding to wavefront aberration before and after compensation is shown in Table 5, and the coefficient after compensation is reduced. This reduction in form error and Zernike coefficients proves the validity and effectiveness of improving optical performance by only compensation the distortion of z-coordinate for cutting tool.

Table 5. Zernike coefficients for wavefront aberration before and after Z-coordinate distortion compensation.

Zernike Item	Before Compensation	After Compensation
q_5	0.356139841	0.355403992
q_6	−0.127976253	−0.127969512

5. Conclusions

Based on the theory of machining optical aspheric surface by FTS turning, the volumetric error modeling and the influence of machining errors are studied in this paper. The effect of the geometric errors on the form accuracy and optical performance is analyzed and simulated, which can make us understand the machining errors effect law in nature. The main conclusions that can be drawn are as follows:

(1) The error components δ_z, ε_{cy} and H_{zx}, as the main machining errors, have an impact on the form accuracy and optical performance of the optical aspheric components, and the contribution of the error δ_z is the largest.

(2) The influence of three main machining error components on wavefront aberration is mainly through increasing the focus and the Y-astigmatism of the optical aspheric surface.

(3) The three main error components affect the form accuracy of machined surface mainly by causing z-coordinate distortion of cutting tool. In the actual process of machining with FTS turning, the compensation in z coordinate of cutting tool will play an active in the form quality and optical performance of the machined components.

Author Contributions: Project administration, J.L.; Supervision, Y.L.; Visualization, X.Z.; Writing—original draft, Y.Z.; Writing—review and editing, A.Y. All authors contributed to writing the manuscript. All authors have read and agreed to the published version of the manuscript.

Funding: This work was financially supported by the National Key R&D Program of China (2016YFE0105100), National Natural Science Foundation of China (61527802), Science and Technology Development Project of Jilin Province (20180201052GX, 20190201303JC, 20191004018TC).

Conflicts of Interest: The authors declare no conflict of interest.

References

1. Cakmakci, O.; Moore, B.; Foroosh, H.; Rolland, J.P. Optimal local shape description for rotationally non-symmetric optical surface design and analysis. *Opt. Express* **2008**, *16*, 1583–1589. [CrossRef]
2. Garrard, K.; Bruegge, T.; Hoffman, J.; Dow, T.; Sohn, A. Design tools for freeform optics. *Proc. SPIE* **2005**, *5874*, 95–105.
3. Jiang, X.; Scott, P.; Whitehouse, D. Freeform surface characterisation-a fresh strategy. *CIRP Ann.* **2007**, *56*, 553–556. [CrossRef]
4. Ramesh, R.; Mannan, M.A.; Poo, A.N. Error compensation in machine tools—A review: Part I: Geometric, cutting-force induced and fixture-dependent errors. *Int. J. Mach. Tools Manuf.* **2000**, *40*, 1235–1256. [CrossRef]
5. Ramesh, R.; Mannan, M.A.; Poo, A.N. Error compensation in machine tools—A review: Part II: Thermal errors. *Int. J. Mach. Tools Manuf.* **2000**, *40*, 1257–1284. [CrossRef]
6. Aikens, D.M. The origin and evolution of the optics for the National Ignition Facility. *Proc. SPIE* **1995**, *2536*, 2–12.
7. Liang, J.C.; Li, H.F.; Yuan, J.X.; Ni, J. A comprehensive error compensation system for correcting geometric, thermal, and cutting force-induced errors. *Int. J. Adv. Manuf. Technol.* **1997**, *13*, 708–712. [CrossRef]
8. Leete, D.L. Automatic compensation of alignment errors in machine tools. *Int. J. Mach. Tool Des. Res.* **1961**, *1*, 293–324. [CrossRef]
9. Ferreira, P.M.; Liu, C.R. An analytical quadratic model for the geometric error of a machine tool. *J. Manuf. Syst.* **1986**, *5*, 51–63. [CrossRef]
10. Okafor, A.C.; Ertekin, Y.M. Derivation of machine tool error models and error compensation procedure for three axes vertical machining center using rigid body kinematics. *Int. J. Mach. Tools Manuf.* **2000**, *40*, 1199–1213. [CrossRef]
11. Lamikiz, A.; de Lacalle, L.N.L.; Ocerin, O.; Diez, D.; Maidagan, E. The Denavit and Hartenberg approach applied to evaluate the consequences in the tool tip position of geometrical errors in five-axis milling centres. *Int. J. Adv. Manuf. Technol.* **2008**, *37*, 122–139. [CrossRef]
12. Kong, L.B.; Cheung, C.F.; To, S.; Lee, W.B.; Du, J.J.; Zhang, Z.J. A kinematics and experimental analysis of form error compensation in ultra-precision machining. *Int. J. Mach. Tools Manuf.* **2008**, *48*, 1408–1419. [CrossRef]
13. Leite, W.D.; Rubio, J.C.C.; Duduch, J.G.; de Almeida, P.E.M. Correcting geometric deviations of CNC Machine-Tools: An approach with Artificial Neural Networks. *Appl. Soft Comput.* **2015**, *36*, 114–124. [CrossRef]
14. Tang, H.; Duan, J.A.; Lan, S.H.; Shui, H.Y. A new geometric error modeling approach for multi-axis system based on stream of variation theory. *Int. J. Mach. Tools Manuf.* **2015**, *92*, 41–51. [CrossRef]

15. Tsutsumi, M.; Saito, A. Identification of angular and positional deviations inherent to 5-axis machining centers with a tilting-rotary table by simultaneous four-axis control movements. *Int. J. Mach. Tools Manuf.* **2004**, *44*, 1333–1342. [CrossRef]

16. Li, D.X.; Feng, P.F.; Zhang, J.F.; Yu, D.W.; Wu, Z.J. An identification method for key geometric errors of machine tool based on matrix differential and experimental test. *Proc. Inst. Mech. Eng. C-J. Mech. Eng. Sci.* **2014**, *228*, 3141–3155. [CrossRef]

17. Cheng, Q.; Zhao, H.W.; Zhang, G.J.; Gu, P.H.; Cai, L.G. An analytical approach for crucial geometric errors identification of multi-axis machine tool based on global sensitivity analysis. *Int. J. Adv. Manuf. Technol.* **2014**, *75*, 107–121. [CrossRef]

18. Liu, X.; Zhang, X.; Fang, F.; Zeng, Z. Performance-controllable manufacture of optical surfaces by ultra-precision machining. *Int. J. Adv. Manuf. Technol.* **2018**, *94*, 4289–4299. [CrossRef]

19. Carmichael Martins, A.; Vohnsen, B. Measuring ocular aberrations sequentially using a digital micromirror device. *Micromachines* **2019**, *10*, 117. [CrossRef]

20. Fuerschbach, K.; Rolland, J.P.; Thompson, K.P. Theory of aberration fields for general optical systems with freeform surfaces. *Opt. Express* **2014**, *22*, 26585–26606. [CrossRef]

21. Liu, T.R.; Qiu, X.; Li, H.; Li, Y.L.; Xu, C.; Hu, Y.; Zhou, X.Q. The simulation study of the influence of surface error on optical properties for optical part. *Adv. Mater. Res.* **2013**, *690*, 3027–3031. [CrossRef]

22. Tamkin, J.M.; Milster, T.D.; Dallas, W. Theory of modulation transfer function artifacts due to mid-spatial-frequency errors and its application to optical tolerancing. *Appl. Opt.* **2010**, *49*, 4825–4835. [CrossRef] [PubMed]

23. Lawson, J.K.; Aikens, D.M.; English, R.E.; Wolfe, C.R. Power spectral density specifications for high-power laser systems. *Proc. SPIE* **1996**, *2775*, 345–356.

24. Lawson, J.K.; English, R.E.; Sacks, R.A.; Trenholme, J.B.; Williams, W.H.; Cotton, C.T. NIF optical specifications: The importance of the RMS gradient. *Proc. SPIE* **1998**, *3492*, 336–343.

25. Gao, H.M.; Fang, F.Z.; Zhang, X.D. Reverse analysis on the geometric errors of ultra-precision machine. *Int. J. Adv. Manuf. Technol.* **2014**, *73*, 1615–1624. [CrossRef]

26. Liu, X.L.; Zhang, X.D.; Fang, F.Z.; Zeng, Z.; Gao, H.M.; Hu, X.T. Influence of machining errors on form errors of microlens arrays in ultra-precision turning. *Int. J. Mach. Tools Manuf.* **2015**, *96*, 80–93. [CrossRef]

27. Zou, X.; Zhao, X.; Li, G.; Li, Z.; Sun, T. Sensitivity analysis using a variance-based method for a three-axis diamond turning machine. *Int. J. Adv. Manuf. Technol.* **2017**, *92*, 4429–4443. [CrossRef]

28. Yang, J.X.; Guan, J.Y.; Ye, X.F.; Li, B.; Cao, Y.L. Effects of geometric and spindle errors on the quality of end turning surface. *J. Zhejiang Univ.-Sci. A* **2015**, *16*, 371–386. [CrossRef]

29. Liu, X.L.; Zhang, X.D.; Fang, F.Z.; Liu, S.G. Identification and compensation of main machining errors on surface form accuracy in ultra-precision diamond turning. *Int. J. Mach. Tools Manuf.* **2016**, *105*, 45–57. [CrossRef]

30. Chen, X. CCD based digital optical transfer function testing instrument. *Proc. SPIE* **2009**. [CrossRef]

micromachines

MDPI

Article

High Temperature Adiabatic Heating in μ-IM Mould Cavities—A Case for Venting Design Solutions

Matthew Tucker, Christian A. Griffiths *, Andrew Rees and Gethin Llewelyn[iD]

College of Engineering, Swansea University, Swansea SA1 8EN, UK; 709390@Swansea.ac.uk (M.T.);
Andrew.Rees@Swansea.ac.uk (A.R.); 656688@swansea.ac.uk (G.L.)
* Correspondence: c.a.griffiths@swansea.ac.uk

Received: 20 February 2020; Accepted: 25 March 2020; Published: 30 March 2020

Abstract: Micro-injection moulding (μ-IM) is a fabrication method that is used to produce miniature parts on a mass production scale. This work investigates how the process parameter settings result in adiabatic heating from gas trapped and rapidly compressed within the mould cavity. The heating of the resident air can result in the diesel effect within the cavity and this can degrade the polymer part in production and lead to damage of the mould. The study uses Autodesk Moldflow to simulate the process and identify accurate boundary conditions to be used in a gas law model to generate an informed prediction of temperatures within the moulding cavity. The results are then compared to physical experiments using the same processing parameters. Findings from the study show that without venting extreme temperature conditions can be present during the filling stage of the process and that venting solutions should be considered when using μ-IM.

Keywords: micro injection moulding; adiabatic heating; diesel effect; venting

1. Introduction

The demand for micro parts has significantly increased due to the ability of the Micro-injection Moulding (μ-IM) manufacturing process to produce a wide variety of components [1,2]. Different μ-IM machines are appearing within the market and these machines can now meet the high accuracy and dimensional requirements demanded by the consumers [3,4]. There is an ever increasing demand for the production of large quantities of micro parts and μ-IM can meet the technical requirements of these products [5].

Previous studies in μ-IM have demonstrated that varying processing conditions such as mould and melt temperature, injection speed and air evacuation in the mould cavity can all have an effect on the resulting process outputs. A study by Griffiths et al., showed that that barrel temperature and injection speed are the key factors that influence the aspect ratios of micro features replicated in Polypropylene (PP) and Acrylonitrile butadiene styrene (ABS) [5]. Further studies have concluded that in μ-IM a high melt and mould temperature along with high injection speeds are required to enhance micro component replication fidelity [6–8]. Mönkkönen et al. found that in addition to the polymer used the most influential parameters were the injection speed, melt temperature and holding pressure [9], whereas others such as Shen et al., found that the most influential parameter was having a mould temperature above the polymer glass transition temperature was optimum for consistent replication of micro injected parts [10]. In a study by Yu et al., the quality of filling as a function of distance from the gate concluded that cavities towards the end of the melt flow filled to a higher yield [8]. The general consensus from the literature is that the high processing setting benefit the μ-IM process [5–10]. Whereas the increase in the process setting of higher temperatures and injection speed is often seen as an optimisation solution for μ-IM, it is clear that there are also negative effects to these process settings. In particular, an increase in the melt and mould temperature increases the process cycle time as extra cooling time is required [5].

During the filling process, the polymer has a specific volume (*V*), which varies depending on both the resultant pressure (*P*) and temperature (*T*). *PVT* data is used to represent the melt flows compressibility and shows the functional dependence between the polymers volume, pressure and temperature [11]. When cavity air temperature is kept at a constant heat for a given volume, the temperature will only rise when the air is further compressed resulting in work performed on the system. This increase in temperature is referred to as the adiabatic temperature and the whole procedure is referred to as adiabatic heating. Furthermore, as the temperature is increased there is an increase in the resulting pressure which will continue to increase the more it is compressed [12]. For the adiabatic heat process, the ideal gas law outlines the relationship between the air Pressure (*P*), Volume (*V*), Temperature (*T*)and the number of moles (*n*) of an ideal gas. For an adiabatic process, there is no heat transfer that takes place, meaning no heat is added or removed to the system. An ideal gas has a given number of atoms in a specified volume and when this changes it inversely effects the pressure and linearly the temperature. This is expressed in Equation (1) below where *R* is the value for the universal gas constant.

$$PV = nRT \tag{1}$$

where for an adiabatic process the final temperature T_2 is given as:

$$T_2 = T_1 \left(\frac{V_1}{V_2} \right)^{\gamma-1} \tag{2}$$

In addition, the final pressure P_2 is given as:

$$P_2 = P_1 \left(\frac{V_1}{V_2} \right)^{\gamma} \tag{3}$$

where T_1, P_1 and V_1 are the initial state values and T_2, P_2 and V_2 are the final state values. γ is the ratio of heat capacity at constant pressure (C_p) to heat capacity at constant volume (C_v) [13]. Air is a diatomic gas made up of around 78% Nitrogen, 20% Oxygen, 0.9% Argon and the further 1.1% made up of additional elements and has a specific heat ratio given as $\gamma = 1.4$ [13].

The adiabatic process can result in extremely high temperatures within the μ-IM mould, with the potential to cause ignition of the compressed gases. This combustion is referred to as the diesel effect, which can lead to damage of the mould cavity as well as degradation of the polymer part [7]. Like the cycle of a diesel engine, the polymer is injected at high speed where it rapidly compresses the air within the mould cavity, which in turns leads the temperature to adiabatically increase.

This high-pressure environment within a mould cavity combined with volatile gases chemically released from the polymer can increase the likelihood of ignition as the auto ignition temperature of the air and gas mixture is achieved. For ignition to occur, there must be a stoichiometric mixture of oxygen and process gasses within the cavity [14]. Process issues due to the combustion in the mould are typically burn marks, short shots, poor surface finish or a change in structural property of the polymer [15]

During the μ-IM process failure to integrate air evacuation within the cycle has the potential to introduce poor process control and damage the mould tooling [12]. When the polymer melt is injected into the mould cavity, the flow front pushes the unvented air towards the end of the cavity causing it to compress. This compression increases the air temperature significantly within the mould. It can result in outgassing, decomposition of mould compounds or leave corrosive residue [12]. Within conventional injection moulding, it is found that the resident air will be evacuated out of the mould cavity through the mould parting line and coarse grain grinding of the surfaces can be used to facilitate this [16]. Passive venting can also be provided with gaps machined at the parting line surfaces (usually at the end of a flow front). These permissible vents allow air to escape without significant pressure build. Usually they are 1.5–2.5 mm wide and the vent depth depends on the polymer being processed. PP, Polyamide (PA), Polyoxymethylene (POM), Polyethylene (PE) use a <15 µm depth vent and Polystyrene (PS), ABS, Polycarbonate (PC), polymethylmethacrylate (PMMA) use a <30 µm depth vent [16]. In addition to this, the pressurised flow front can push air into ejector pins. Venting pins can be made if the pin is 20–50 µm smaller than the bore for a length of 300 µm [16].

Often the dimensions of these vents are larger than some micro parts, so macro venting design rules do not easily translate to micro moulds. Yao and Kim identified that components manufactured by µ-IM fall into one of the following two categories. Type A are components with overall sizes of less than 1 mm while Type B have larger overall dimensions but incorporate micro features with sizes typically smaller than 200 µm [17]. So it can be seen that due to scaling issues some of the rules for macro venting solutions cannot be used in µ-IM. Ideally, the primary vent is present at the split line of the mould faces but with high precision surface achieved when manufacturing moulds using micro machining processes [18] there can be insufficient gaps between the split lines.

Currently, changes to the processing conditions and altering the injection locations and injection speed profiles is used to prevent air traps but with very fast injection times this method has limitations. Also, the majority of micro parts are considered to be 'blind-holes' where the air gets trapped and gets compressed at high flow speeds and this cause resistance to the melt flow resulting in abnormalities within the final part production such as uneven flow fronts [19].

To aid the evacuation of air from a mould cavity, vacuum venting has been introduced within µ-IM tooling platforms. This feature aids in the counteracting pressure that is being produced by the compressed air at the flow front which improves replication accuracy and process control [20]. Utilising vacuum venting results in the ability to reduce the flow resistance of the polymer melt that would usually be effected by trapped air within the cavity or micro features [21]. In the study by Lucchetta et al., the importance of ensuring that assisted venting does not reduce the temperature of the mould surface was highlighted for temperature sensitive polymers [22].

This research will demonstrate the requirement for venting design solutions when considering mass manufacture of polymer micro components. A micro test part will be produced using the µ-IM process to establish if adiabatic heating will occur from gas trapped and rapidly compressed within the mould cavity. In the methodology section, the Autodesk Moldflow (Autodesk, San Rafael, CA, USA) setup for simulating part manufacture with varying processing parameters using the Design of Experiments (DOE) method is shown. This is then followed by the application of an adiabatic heating model using the boundary conditions established from the simulation to generate an informed prediction of temperatures within the moulding cavity. Finally, the adiabatic heating and diesel effect results are presented and compared to parts produced with the same process settings used in the simulation.

2. Experimental Procedure

2.1. Part Design

For this study, a geometry (Figure 1) was used which is suitable for replication using a Battenfeld Microsystem 50 µ-IM machine (Wittmann Battenfeld GmbH, Kottingbrunn, Austria). The geometry had a constant thickness of 0.5 mm, a runner length of 40 mm and a final rectangular section measuring 15 mm × 5 mm. Each of the corners had a radius of 0.5 mm in order to reduce shear. The geometry had an overall surface area of 249.75 mm^2 and a volume of 47.61 mm^3. The total flow length of the part was 55.8 mm (Table 1).

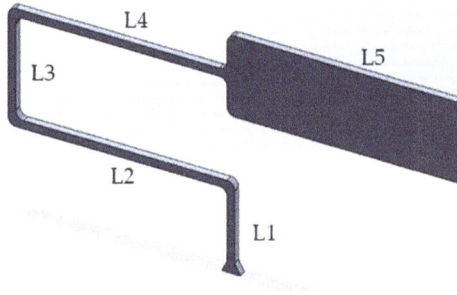

Figure 1. Test part geometry.

Table 1. Section dimensions for the test part.

Length	L1	L2	L3	L4	L5
Distance (mm)	5	14.5	7.3	14	15
Length sum (mm)	5	19.5	26.8	40.8	55.8

2.2. Moldflow Simulation

In this study, the simulation software Autodesk Moldflow Insight 2018 was used. A study by Xie et al. concluded that a 2.5D meshes such as the Moldflow dual domain mesh do not capture all effects taking place [23]. Therefore, a full 3D meshwais used. Following preliminary mesh sensitivity analysis an element size of 0.1 mm was used as illustrated in Figure 2.

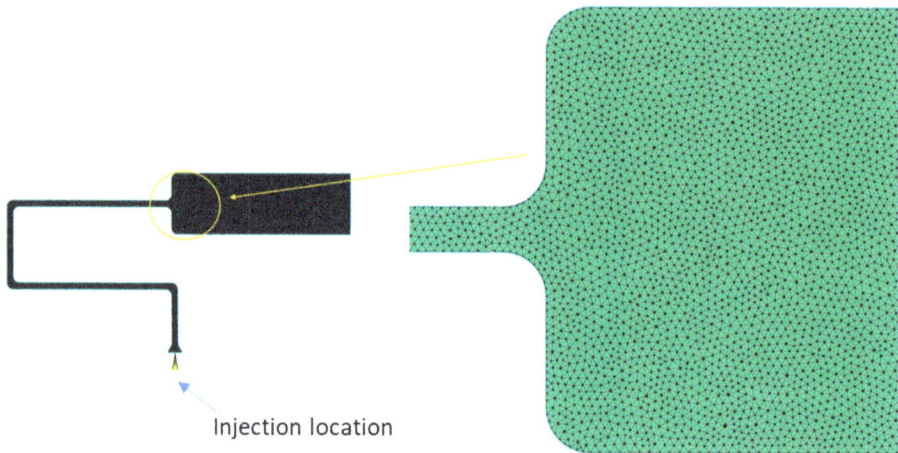

Figure 2. Element size 0.1mm 3D (Fine mesh).

2.3. Materials

The material properties for PP and ABS are displayed in Table 2. Both materials are used extensively within an industrial context [24].

Table 2. Material properties of Polypropylene (PP) and Acrylonitrile butadiene styrene (ABS).

Description	(PP)	(ABS)
Family Name	Polypropylenes (PP)	Acrylonitrile Copolymers
Trade Name	SABIC PP 56M10	MAGNUM 8434
Manufacturer	SABIC Europe B.V.	Trinseo EUR
Moldflow Viscosity Index	VI(240)0087	VI(240)0212
Transition Temperature °C	150	50
Specific Heat Data		
Temperature °C	240	240
Specific Heat (Cp) J/kg·°C	2750	2032
Thermal Conductivity Data		
Temperature °C	240	240
Thermal Conductivity W/m·°C	0.18	0.152
Mechanical Properties		
Elastic Modulus	1340 MPa	2240 MPa
Poisson Ratio	0.392	0.392
Shear Modulus	481.3 MPa	804.6 MPa
Environmental Impact		
Resin ID code	5	7
Energy Usage Indicator	3	5

2.4. Boundary Conditions

The simulation boundary conditions and process settings were the same as those used when processing PP and ABS on the Battenfeld 50 µ-IM machine. The mould geometry was set as solid walls whilst the tooling material was P-20 tool steel. Table 3 below shows the mechanical and thermal properties of the steel. One injection gate was used and is displayed in Figure 2. Table 4 illustrates the process variables used for the simulation L9 DOE.

Table 3. Mechanical and thermal properties of P-20 tool steel.

Mould Specific Heat	460 J/Kg·°C
Mould Thermal conductivity	29 W/m·°C
Elastic Modulus	205,000 MPa
Poisson ratio	0.29

Table 4. Test parameters.

Test No.	Melt Temp (°C)		Mould Temp (°C)		Injection Speed (mm/s)	
	PP	ABS	PP	ABS	PP	ABS
Test 1	220	220	20	40	200	200
Test 2	250	250	40	60	500	500
Test 3	270	280	60	80	800	800
Test 4	220	220	40	60	800	800
Test 5	250	250	60	80	200	200
Test 6	270	280	20	40	500	500
Test 7	220	220	60	80	500	500
Test 8	250	250	20	40	800	800
Test 9	270	280	40	60	200	200

3. Results

3.1. Simulation

3.1.1. Moldflow Air Trap Results

Air traps are formed when the flow front of the polymer melt compresses air against the cavity wall when there is insufficient venting in place. The simulations identify regions of the test part where air is trapped and where there is potential for the diesel effect. For PP all nine tests successfully filled the mould and all contain air traps on the end wall of the cavity. Figure 3 shows an air trap witnessed in test one and this is a typical representation from all performed experiments. The result is informative in that it shows the part and tool designer where potential part quality issues may arise due to gassing and where vents need to be considered. When processing with ABS only three tests which utilised the higher melt temperatures (Test 3, 6 and 9) achieved a completely filled cavity. For the filled parts the air trap positions are similar to those observed for the PP simulations.

Figure 3. Air traps witnessed in Polypropylene (PP) simulations.

3.1.2. Moldflow PP and ABS Temperature Results

Figure 4 displays the resulting Flow Front Temperatures (T_{ff}) reached in the simulations when compared against the input Temperatures (T_i) detailed in Table 4. For both materials an increase in T_{ff} temperature from the T_i is observed. Test 4 had the largest increase in temperatures from the T_i. In particular, for PP and ABS an increase in 8.13% and 19.5% respectively was witnessed.

Typically, polymers have an absolute maximum melt temperature. Any increase in this value can result in degradation of the polymer. For the PP test 3 experiment, the absolute maximum melt temperature for the material (280 °C) is exceeded by 2.6 °C due to a shear heating temperature in the cavity. This means that this setting should be avoided and that there is also potential for increased gassing from polymer degradation. For both materials it can be seen that the process window based on the four factors has a large influence on the temperature of the polymer within the cavity and this has the potential to influence the temperature of the resident air within the cavity.

3.1.3. Moldflow PP and ABS Shear Rate Results

In Figure 5, the simulation shear rate results for PP and ABS are presented. For both materials, tests three, four and eight had the largest presence of shear within their cycles. These three tests utilised the highest injection speed of 800 mm/s. In comparison tests one, five and nine achieved the lowest presence of shear having been processed at the lowest injection speeds of 200 mm/s. The findings demonstrate that the process parameters have a major influence on the resulting shear but importantly no single test goes above the critical shear rate for PP (1×10^5 1/S) and ABS (5×10^4 1/S) [25]. The results demonstrate the potential of shear to increase the temperature and contribute to the formation of gasses within the moulding cavity.

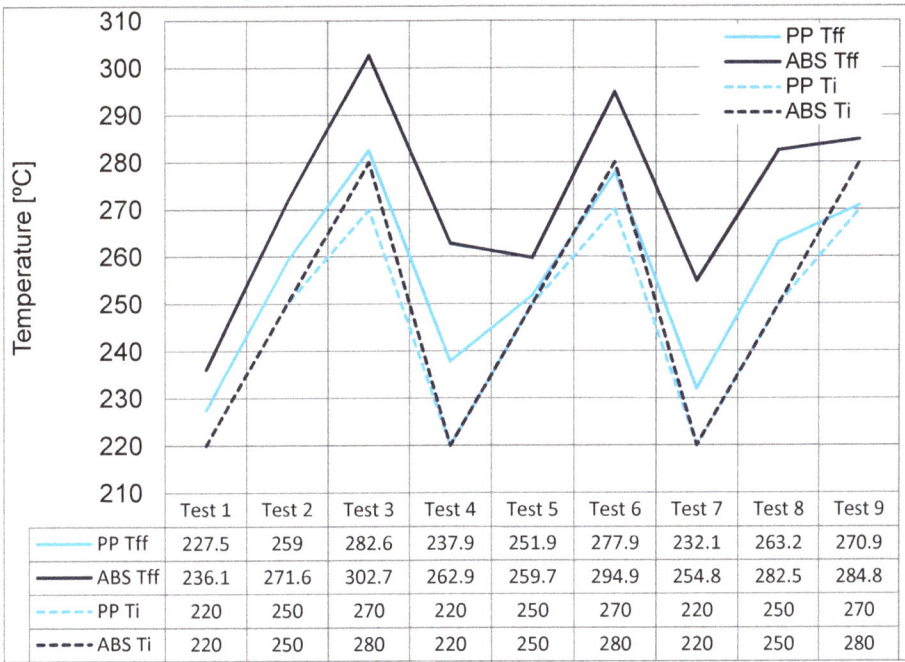

	Test 1	Test 2	Test 3	Test 4	Test 5	Test 6	Test 7	Test 8	Test 9
PP Tff	227.5	259	282.6	237.9	251.9	277.9	232.1	263.2	270.9
ABS Tff	236.1	271.6	302.7	262.9	259.7	294.9	254.8	282.5	284.8
PP Ti	220	250	270	220	250	270	220	250	270
ABS Ti	220	250	280	220	250	280	220	250	280

Figure 4. Comparison between T_i and T_{ff} on filling for PP and Acrylonitrile butadiene styrene (ABS).

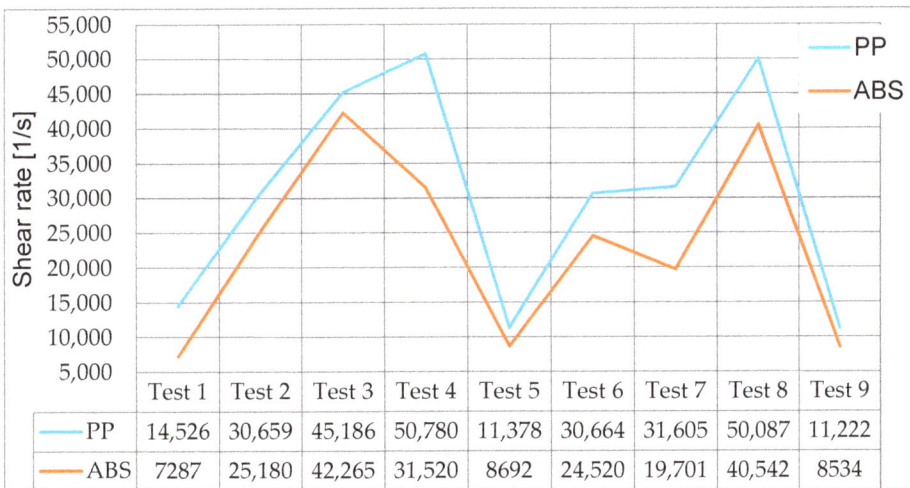

	Test 1	Test 2	Test 3	Test 4	Test 5	Test 6	Test 7	Test 8	Test 9
PP	14,526	30,659	45,186	50,780	11,378	30,664	31,605	50,087	11,222
ABS	7287	25,180	42,265	31,520	8692	24,520	19,701	40,542	8534

Figure 5. Maximum shear rate for PP and ABS.

3.1.4. PP Analysis of Individual Tests

Further analysis has been performed using the results presented in Figure 4. In particular, test three where the highest bulk and T_{ff} have been observed. Bulk temperature is used, as it is too difficult to display the temperature change of the polymer within one display as the temperature profile is dynamic and changes with time, location and thickness during the injection process. It also represents the energy transported through particular locations in which it has more physical significance than average temperature. The T_{ff} value is obtained via the fill analysis within Moldflow and represents the polymer temperature as it reaches a specified point. The original melt temperature T_i at the point of injection for test three was 270 °C (T_i) and it increased to 286 °C (T_{ff}) for the bulk temperature. Also, the flow front temperature had an increase from 270 °C (T_i) to 282.6 °C (T_{ff}). Within the Moldflow analysis the largest region of temperature increase is within the runner of the moulded part where the polymer is experiencing shear, then when it flows into the main cavity it starts to cool. Moldflow does not account for adiabatic heating of the resident air within the model hence no further temperature increases are observed. In reality the flow conditions within the mould presents an ideal environment for the diesel effect to occur whereby the resident air would see a rapid rise in initial temperature before adiabatically compressed.

3.1.5. ABS Analysis of Individual Tests

As with the comparison of the temperature results for PP, test three will be used as this has the highest process parameter conditions and is the most likely test that would see the diesel effect occur. The original melt temperature T_i at the point of injection was 280 °C (T_i) and it increased to 303.4 °C (T_{ff}) for the bulk temperature. Also, the flow front temperature increased from 280 °C (T_i) to 302.7 °C (T_{ff}). This temperature rise is the consequence of shear heating as observed in Figure 5.

3.2. Adiabatic Heating

3.2.1. Adiabatic Conditions for PP

As the Moldflow software does not account for compressional heat rise in the model, further analysis is required to understand the temperature and shear rate simulation results. Using the results of the T_{ff} from Moldflow together with the ideal gas law for adiabatic heating (Equation (1)), it is possible to estimate the temperature of the compressed air within the mould cavity. Figure 6 shows the pressure conditions that lead to adiabatic heating occurring for both the PP and ABS polymers. It can be seen that the pressure increases in accordance to the reduction of the volume within the mould cavity. In particular of major relevance for μ-IM is that when the unfilled cavity volume is reduced beyond 5 mm³ the pressure gradients are altered significantly and there is an exponential increase of pressure in the mould cavity. This shows that without venting of the air within the cavity there is an adiabatic process that is taking place, which can result in diesel effect occurring within the mould.

If the temperature of the resident air within the cavity is already equal to the mould temperature, it can be expected that the resident air will heat up rapidly when in contact with the flow front temperature. Without any venting, the rapid compression of this heated air during the filling process will result in a diesel effect where temperatures within the mould will rise to above 1300 °C (Figure 7). The model calculates that the resident air is compressed to around 0.5 mm³, which simulates an air trap in the mould cavity. According to the ideal gas law the smaller the volume that is achieved under compression the higher the final temperature and it is expected that combustion will occur before such extreme temperatures are reached. At high temperature gassing has already occurred in which fumes are released from the chemical properties of the polymers; this is potentially harmful to the mould and localised etching can be expected. Ignition of the resident gas can take place above 357 °C, as this is the flashpoint of PP as marked by the blue dashed line on the graph in Figure 7. Above this line the risk of damage to the part and the mould remain high. As shown, combustion will occur when the resident air is compressed to around 15 mm³. However, it can be seen that the air-gas mixture has the potential to reach higher temperatures as the volume of air is reduced.

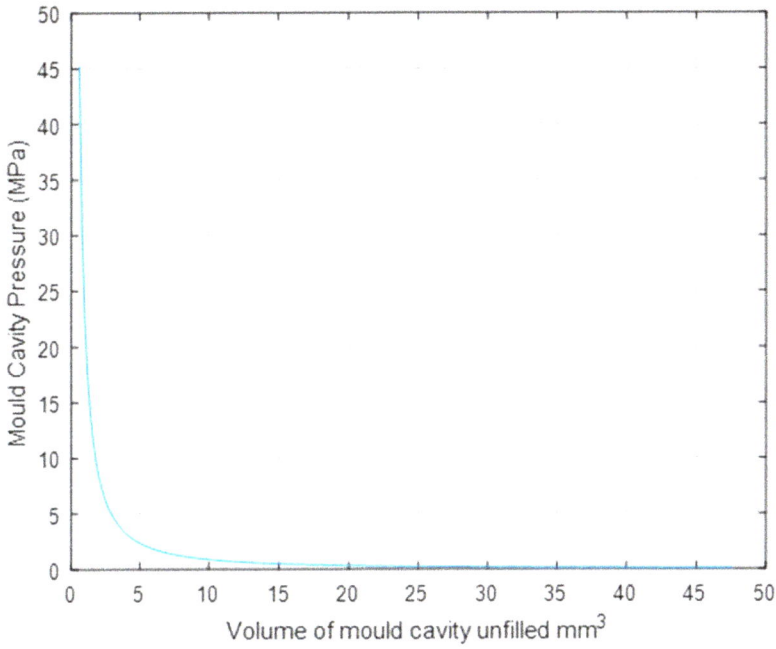

Figure 6. Pressure increase when air is compressed in mould.

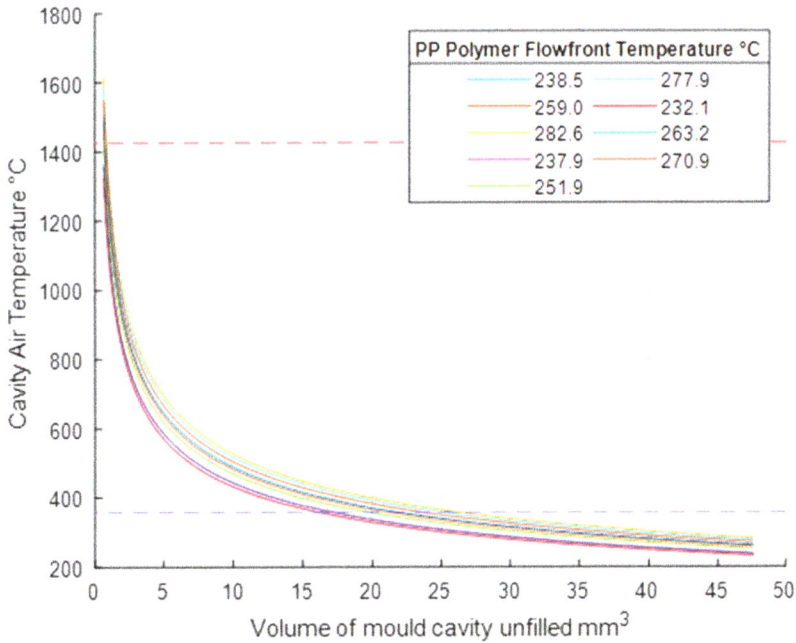

Figure 7. Adiabatic temperature increase when air is compressed in mould, PP.

The results in Figure 7 display extremely high temperatures that without adequate venting could be experienced for all nine tests. This can cause extreme problems for the mould cavity itself. To demonstrate this, Figure 7 also displays a red horizontal dashed line indicating the melting temperature of the P20 steel at 1426 °C. The continuous cyclic temperature rise and fall has the potential of causing damage to the mould steel. The potential localised microstructure change could compromise the tooling integrity as typically clamping forces for the mould can reach 50 kN [26]. Previous research studies have demonstrated that venting has proven to reduce the resulting cavity pressure [27]. In addition, studies have demonstrated that when processing PP above the recommended melt temperatures of 180 °C to 240 °C in non-vented moulds there can be severe chemical degradation within the polymer part brought on by elevated temperatures [28].

3.2.2. Adiabatic Conditions for ABS

Similar to the results for PP, the ABS polymer also undergoes adiabatic heating according to the ideal gas law. Assuming that the polymer T_{ff} increases the resident air temperature, Figure 8 shows the temperature profile when adiabatic heating occurs. All test results show a resulting cavity temperature above 1300 °C. It is noted that the T_{ff} received from the ABS results in a higher adiabatic temperature increase when compared to PP. In Figure 8 the red dashed lines indicates the melt point temperature of 1426 °C for the P20 tool steel mould material [29]. With the addition of adiabatic temperature rises, mould temperatures can exceed the auto ignition temperature of ABS which is between 500 °C and 575 °C and represented by the blue dashed line in Figure 8. This demonstrates that without venting all experiments can produce the diesel effect and lead to part and mould damage.

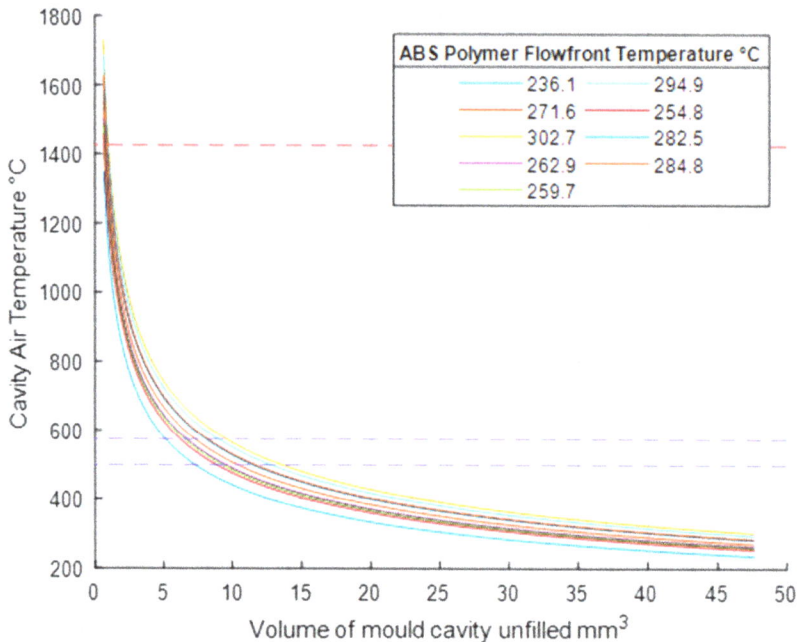

Figure 8. Adiabatic temperature increase when air is compressed in mould, ABS.

3.3. Comparison with Experimental Data

In order to validate the simulation results, the experimental work that Griffiths et al. presented is used. This work uses the same part (Figure 1) and process settings (Table 4) used for the simulation results. 180 test runs were performed and the part flow lengths of ten parts for each process combination were measured. For the simulations the Moldflow measurement tool is used to measure the part flow length and it is calculated from the beginning of the runner to the end of the flow front. The physical parts produced had irregular flow fronts so the difference between the highest and lowest portion of the flow front was measured for each part.

It was found that the tests with the highest process parameters had the largest flow lengths (Table 5) [30]. The experimental tests revealed that when comparing the two polymers, PP had the highest average flow length for all nine of the tests conducted. Not all of the ABS parts filled so it was not possible to produce a trapped air and diesel effect in the cavity, therefore these experiments are not considered. For the PP physical experiments, tests three, six and nine had the highest melt temperature and achieved the best filling results (Figure 9). Figure 9 also shows the end of fill images imposed on the CAD model for test one, seven and three. Test three had the highest flow length and this was achieved with high melt and mould temperatures and the highest injection speed of 800 mm/s. This supports the simulation results, which show that these tests can reach the highest T_{ff} (Figure 4). However, the simulation could not show the trapped air pocket as the simulation assumes the mould is fully vented and therefore no air is trapped.

Table 5. Flow lengths for tests of PP and ABS [30].

Test Number	Flow Length PP (mm)	Flow Length ABS (mm)
Test 1	47.2	32.6
Test 2	52	42.4
Test 3	54.7	45
Test 4	48.9	39.8
Test 5	52.8	32.5
Test 6	54.6	29.3
Test 7	49.7	32.8
Test 8	52.7	36.4
Test 9	54.1	33.1

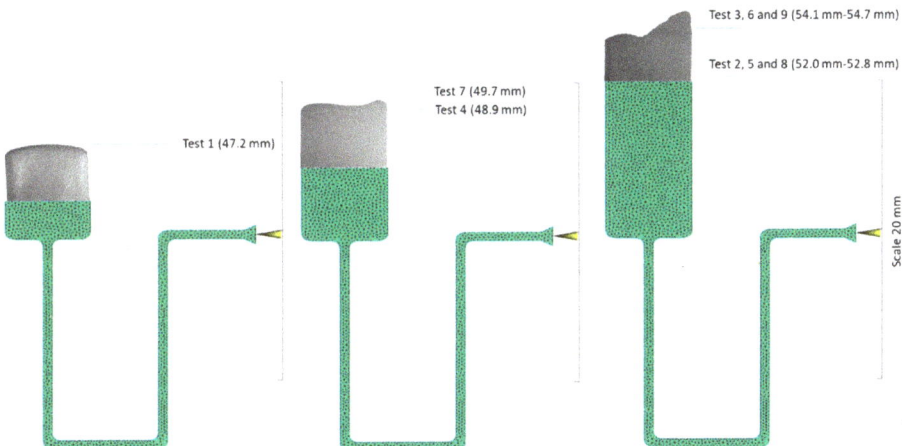

Figure 9. PP flow lengths for physical Tests 1-9.

The adiabatic model used in this research proves that it is possible to produce a diesel effect if the mould used does not have sufficient venting. The physical experiments also show that the diesel effect is present. The PP results show that the experiment with the lowest flow length (Table 5) and the lowest T_{ff} (Figure 4) was test 1. It can be seen in Figure 10 that the part has a rounded flow front as expected for a part that is unfilled. Because the part is unfilled there is no possibility of an excessive gas trap. For test three in Figure 9 there is clear evidence of a gas trap. This experiment has the highest flow length and the highest T_{ff} and as seen in the figure the flow front is uneven and uncharacteristic of a normal flow front. On further inspection of test part from this experiment it can be seen that there are gas pockets within the parts and there is evidence of polymer damage (Figure 11). The PP test parts produced with a high melt temperature displayed signs of part damage due to the resident air inside the cavity, this result confirms the high temperatures observed in the simulations.

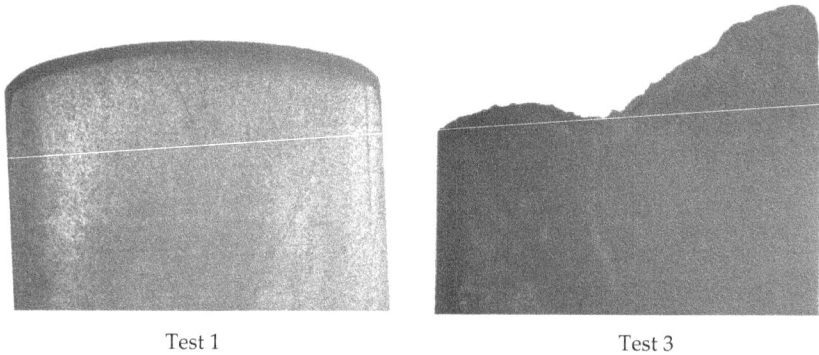

Test 1   Test 3

Figure 10. Flow fronts for test parts 1 and 3.

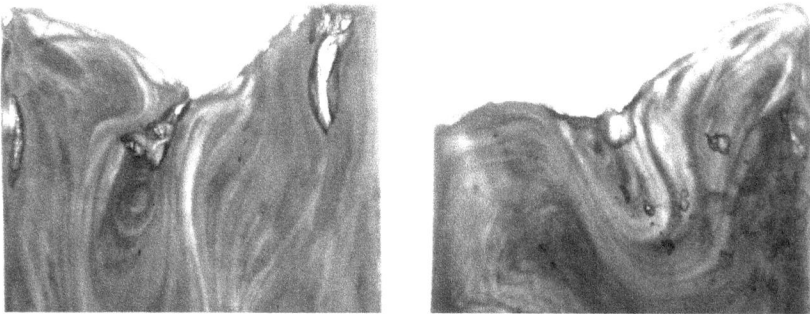

Figure 11. Test part 3 flow fronts with gas trapped within the parts.

4. Conclusions

This paper presents an investigation on the influence on μ-IM process parameter settings on adiabatic heating from gas trapped and rapidly compressed within the mould cavity. In the study PP and ABS polymers were processed with a range of different parameters to identify boundary conditions for use within a gas law model. The model is then used to identify adiabatic conditions within the mould cavity. The conclusions are as follows:

- Autodesk Moldflow simulations of an established part design can predict accurate temperature distributions within the μ-IM process. These results can then be used to identify accurate boundary conditions to be used in the gas law model to generate an informed prediction of temperature increases within the moulding cavity.
- In a mould with limited venting, extreme temperature conditions can be present during the filling stage of the process. The results show the maximum air temperature while processing PP can exceed 1300 °C when the melt flow front temperatures are between 238–283 °C. When processing ABS material the mould temperature can exceed 1400 °C when the melt flow front temperatures

are above 270 °C. With such significant temperature increases it is highly likely that the polymer parts will degrade and the tooling will experience damage with prolonged use.

- Further work should consider improving the model by the addition of the heat transfer rate of the polymer flow front temperature to the resident air within the mould cavity. The influence of different polymer gases during processing should also be considered for their contribution to the diesel effect.

- The simulation of the factors that influence temperature together with the gas model highlight the potential for adiabatic heating and the physical experiments show that gas traps and part damage are experienced with combinations of process settings. The model shows extreme temperatures within the cavity, the highest temperatures are unlikely to arise as there will always be some natural venting. However, it also shows that with limited venting there is a temperature increase that is detrimental to the process. Due to size limitations macro mould venting solutions cannot always be considered for micro moulds and the findings highlight the need for designs that consider novel venting and air evacuation solutions for improved part quality and tool life.

Author Contributions: For this research conceptualization of the investigation was by C.A.G., the methodology and data curation by C.A.G., M.T. Validation was performed by C.A.G., M.T.; formal analysis and the identification of resources was by A.R. The original draft preparation was done by M.T. and the writing—review and editing was performed by G.L. All authors have read and agreed to the published version of the manuscript.

Funding: This research was funded by the Future Manufacturing Research Institute, College of Engineering, Swansea University and Advanced Sustainable Manufacturing Technologies (ASTUTE 2022) project, which is partly funded from the EU's European Regional Development Fund through the Welsh European Funding Office, in enabling the research upon which this paper is based. Further information on ASTUTE can be found at www.astutewales.com.

Conflicts of Interest: The authors declare no conflict of interest.

References

1. Michaeli, W.; Opfermann, D.; Kamps, T. Advances in micro assembly injection moulding for use in medical systems. *Int. J. Adv. Manuf. Technol.* **2007**, *33*, 206–211. [CrossRef]

2. Muanchan, P.; Kaneda, R.; Ito, H. Polymer Material Structure and properties in Micro injection moulding parts. In *Micro Injection Moulding*; Tosello, G., Ed.; Hanser: Munich, Germany, 2018; pp. 58–81.

3. Fantoni, G.; De Grave, A.; Hansen, H.N. Functional analysis for micro mechanical systems. *Int. J. Des. Eng.* **2010**, *3*, 355–373. [CrossRef]

4. Fantoni, G.; Gabelloni, D.; Tosello, G.; Hansen, H.N. Micro Injection Molding Machines Technology. In *Micro Injection Molding*; Tosello, G., Ed.; Carl Hanser Verlag: Munich, Germany, 2018; pp. 16–17.

5. Sha, B.; Dimov, S.; Griffiths, C.; Packianather, M.S. Micro-injection moulding: Factors affecting the achievable aspect ratios. *Int. J. Adv. Manuf. Technol.* **2006**, *33*, 147–156. [CrossRef]

6. Zhao, J.; Mayes, R.H.; Chen, G.; Xie, H.; Chan, P.S. Effects of process parameters on the micro molding process. *Polym. Eng. Sci.* **2003**, *43*, 1542–1554. [CrossRef]

7. Giboz, J.; Copponnex, T.; Mele, P. Microinjection molding of thermoplastic polymers: a review. *J. Micromech. Microeng.* **2007**, *17*, R96–R109. [CrossRef]

8. Yu, L.; Koh, C.; Lee, J.; Koelling, K.; Madou, M. Experiment investigation and numerical simulation of injection molding with micro-features. *Polym. Eng. Sci.* **2002**, *42*, 871–888. [CrossRef]

9. Mönkkönen, K.; Pakkanen, T.T.; Hietala, J.; Pääkkönen, E.J.; Pääkkönen, P.; Jääskeläinen, T.; Kaikuranta, T. Replication of sub-micron features using amorphous thermoplastics. *Polym. Eng. Sci.* **2002**, *42*, 1600–1608. [CrossRef]

10. Shen, Y.; Wu, W. An analysis of the three-dimensional micro-injection molding. *Int. Commun. Heat Mass Transf.* **2002**, *29*, 423–431. [CrossRef]

11. Chang, R.Y.; Chen, C.H.; Su, K.S. Modifying the tait equation with cooling-rate effects to predict the pressure-volume-temperature behaviors of amorphous polymers: Modeling and experiments. *Polym. Eng. Sci.* **1996**, *36*, 1737–1846.

12. Griffiths, C.A.; Dimov, S.S.; Scholz, S.G.; Tosello, G. Cavity Air Flow Behavior During Filling in Microinjection Molding. *J. Manuf. Sci. Eng.* **2011**, *133*, 11006. [CrossRef]

13. Helmenstine, A. The Chemical Composition of Air. ThoughtCo, 7 July 2019. Available online: https://www.thoughtco.com/chemical-composition-of-air-604288 (accessed on 7 September 2019).

14. Moran, M.; Shapiro, H.; Boettner, D.; Bailey, M. Ideal Gas Mixture and Psychrometric Applications. In *Fundamentals of Engineering Thermodynamics*; Wiley: Hoboken, NJ, USA, 2010; pp. 713–804.

15. Košík, M.; Likavčan, L.; Bilik, J.; Martinkovič, M. Diesel Effect Problem Solving During Injection Moulding. *Res. Pap. Fac. Mater. Sci. Technol. Slovak Univ. Technol.* **2014**, *22*, 97–102. [CrossRef]

16. Menges, G.; Michaeli, W.; Mohren, P. *How to Make Injection Molds*; Carl Hanser Verlag GmbH & Co. KG: Munich, Germany, 2001; p. 631.

17. Yao, N.; Kim, B. Scaling Issues in Miniaturization of Injection Molded Parts. *J. Manuf. Sci. Eng.* **2004**, *126*, 733–739. [CrossRef]

18. Masuzawa, T. State of the Art of Micromachining. *CIRP Ann.* **2000**, *49*, 473–488. [CrossRef]

19. Piotter, V.; Hanemann, T.; Ruprecht, R.; Haußelt, J. Injection molding and related techniques for fabrication of microstructures. *Microsyst. Technol.* **1997**, *3*, 129–133. [CrossRef]

20. Sorgato, M.; Babenko, M.; Lucchetta, G.; Whiteside, B. Investigation of the influence of vacuum venting on mould surface temperature in micro injection moulding. *Int. J. Adv. Manuf. Technol.* **2016**, *88*, 547–555. [CrossRef]

21. Despa, M.S.; Kelly, K.W.; Collier, J.R. Injection molding of polymeric LIGA HARMs. *Microsyst. Technol.* **1999**, *6*, 60–66. [CrossRef]

22. Lucchetta, G.; Sorgato, M.; Carmignato, S.; Savio, E. Investigating the technological limits of micro-injection molding in replicating high aspect ratio micro-structured surfaces. *CIRP Ann.* **2014**, *63*, 521–524. [CrossRef]

23. Xie, L.; Shen, L.; Jiang, B. Modelling and Simulation for Micro Injection Molding Process. In *Computational Fluid Dynamics Technologies and Applications*; InTech: Rijeka, Croatia, 2011; pp. 317–332.

24. Crawford, C.; Quinn, B. Physiochemical properties. In *Microplastic Pollutants*; Elsevier: Amsterdam, The Netherlands, 2017; pp. 57–101.

25. Autodesk Moldflow, "Shear Rate Result," Autodesk Moldflow, 28 May 2017. Available online: https://knowledge.autodesk.com/support/moldflow-insight/learnexplore/caas/CloudHelp/cloudhelp/2018/ENU/MoldflowInsight/files/GUID9BEF20CA-9000-44A9-B572-8F4590F09150-htm.html (accessed on 15 September 2019).

26. Zhang, P.; Xie, P.; Wang, J.; Ding, Y.; Yang, W. Development of a multimicroinjection molding system for thermoplastic polymer. *Polym. Eng. Sci.* **2012**, *52*, 2237–2244. [CrossRef]

27. Rompetrol, S.C. PETROCHEMICALS, "MSDS-01: -POLYPROPYLENE -MATERIAL SAFETY DATA SHEET," 7 May 2008. Available online: https://www.petrobul-bg.com/files/MSDS%20PP%20eng.pdf (accessed on 23 September 2019).

28. White, J. *Principles of Polymer Engineering Rheology*; Wiley-Interscience Publication: Arkon, OH, USA, 1990.

29. Metal Suppliers Online: Material Property Data, Metal Suppliers Online. 2019. Available online: https://www.suppliersonline.com/propertypages/P20.asp (accessed on 12 September 2019).

30. Griffiths, C.; Dimov, S.; Brousseau, E.; Hoyle, R. The effects of tool surface quality in micro-injection moulding. *J. Mater. Process. Technol.* **2007**, *189*, 418–427. [CrossRef]

micromachines

MDPI

Article

Microinjection Molding of Out-of-Plane Bistable Mechanisms

Wook-Bae Kim [1,*] and Sol-Yi Han [2]

[1]   Department of Mechanical Design Engineering, Korea Polytechnic University, Siheung 15073, Korea
[2]   R&D Devision, Eosystem, Incheon 22829, Korea; hansolyi307@gmail.com
[*]   Correspondence: wkim@kpu.ac.kr; Tel.: +82-31-8041-0430

Received: 19 December 2019; Accepted: 28 January 2020; Published: 30 January 2020

Abstract: We present a novel fabrication technique of a miniaturized out-of-plane compliant bistable mechanism (OBM) by microinjection molding (MM) and assembling. OBMs are mostly in-plane monolithic devices containing delicate elastic elements fabricated in metal, plastic, or by a microelectromechanical system (MEMS) process. The proposed technique is based on stacking two out-of-plane V-beam structures obtained by mold fabrication and MM of thermoplastic polyacetal resin (POM) and joining their centers and outer frames to construct a double V-beam structure. A copper alloy mold insert was machined with the sectional dimensions of the V-beam cavities. Next, the insert was re-machined to reduce dimensional errors caused by part shrinkage. The V-beam structure was injection-molded at a high temperature. Gradually elongated short-shots were obtained by increasing pressure, showing the symmetrical melt filling through the V-beam cavities. The as-molded structure was buckled elastically by an external-force load but showed a monostable behavior because of a higher unconstrained buckling mode. The double V-beam device assembled with two single-molded structures shows clear bistability. The experimental force-displacement curve of the molded structure is presented for examination. This work can potentially contribute to the fabrication of architected materials with periodic assembly of the plastic bistable mechanism for diverse functionalities, such as energy absorption and shape morphing.

Keywords: bistable mechanism; V-beam structure; compliant mechanism; microinjection molding; out-of-plane

1. Introduction

Bistable mechanisms (BMs) are defined as mechanical systems that exhibit two stable states in two different positions [1,2]. A movable structure in BMs switches quickly from one position to another when a force is exerted beyond a threshold value and keeps a stable state even under small environmental disturbances without an external power. Due to their unique force-displacement behavior, BMs have been used for diverse applications such as switches, latches, valves, clamps, actuators, robotics, and energy harvesting [3–8].

Unlike the traditional latch-lock, a compliant mechanism produces bistability by storing and releasing strain energy from its flexible structural members during movement. In many microsystems, a compliant bistable (CB) mechanism has been widely used because devices are easily fabricated in monolithic form with no conventional mechanical elements to be assembled, such as bearing, pin and spring. The simplest elastic-buckling CB structure is a bent beam or plate created by simply holding a business card between two fingers and folding it. It snaps laterally and stays this way after applying an external force to its surface.

To create this buckling behavior on a small-scale structure, the axial load necessary may result from the fabrication process residual stress or from direct compression of the corresponding beam with a come drive or electrothermal actuator conveniently placed on it [9,10]. A more simplified method to obtain buckling motion without residual stress and stress loading is to use double curved or V-shaped beams clamped together at their centers [11–13].

In microsystems, BMs use mainly silicon-based materials, bringing benefits such as compliant structure precision, process reproducibility, and high mechanical strength. However, silicon materials, as compared to polymer and metal, suffer from compliant-motion limited displacement due to their small yield strain (Ratio of strength to Young's modulus). Metal offers good properties to BMs for its strength and toughness. However, it has limited miniaturization and a high manufacturing cost, the latter due to elaborate fabrication processes such as wire electric discharge machining (WEDM) and precision cutting. On the other hand, MEMS fabrication, using metal and silicon materials, need a complex process to produce three-dimensional structures [14,15].

Plastic is a good candidate for compliant mechanisms (CMs) including bistable systems due to its flexibility and cost effectiveness. Its yield strain is in the order of 5% to 10%, higher than that of metal, silicon, and ceramics (in metals it is close to 0.1% [16]). Although plastic is less resistant and its behavior unpredictable, it is essentially useful for structures with low stiffness and large displacement.

An injection molding process (IM) can offer the freedom to create three-dimensional structures WEDM and MEMS cannot provide; it offers mass production at a relatively low cost, making it suitable for micro-CMs. Nevertheless, there are typical problems with monolithic CMs injection molding. It is a challenge to apply it to an elastic member like a slender beam, as the small cavity thickness caused by short shot causes the quick cooling of the molten polymer. This is common in the MM process for various high precision micro-components such as micro-gear, microfluidic devices, and microlens arrays [17,18]. For more than ten years, there has been a growing interest in MM, especially on the influence of replication-fidelity process parameters such as melting and molding temperature, injection speed and holding pressure. High settings of these parameters generally give a positive effect on the replication of micro-features. In addition, the influence degree of key parameters could depend on part layout, materials, and mold roughness and coating features [19–22].

Plastic micro-cantilever beams have been injection-molded into micro-chemical and biomedical sensors, atomic force microscopes, and micro-springs. Injection molding of micro-cantilever or micro-bridge structures were the subject of several studies that seek to understand the mechanism of microinjection molding [23–28]. A few of those reveal a high length-to-thickness ratio in beam structures in the order of tens or even over a hundred. This is possible using a mold temperature controller that switches to a high setting in a melt-filling stage and to low in a cooling one. Except for a micro-cantilever, there has been little research on good quality, high length-to-thickness ratio beams for specific function; that is, fabrication and testing of compliant BMs using MM, and as far as the authors know, it has hardly ever been researched.

In this study, stacking two single V-beam structures to construct the double V-beam structure is proposed as a new method to produce OBMs. We first designed out-of-plane BMs using 150 μm thick vertically inclined slender beams, forming V-shape structures. We examined the part layout for IM success of a single V-beam structure and observed the force-displacement behavior of the stacked double V-beam structure dimension through a finite element analysis (FEA). Then, IM was used to create the designed single mechanisms and molding parameters were set to obtain parts without defects, such as flashes and short shots. Dimensional errors caused by shrinkage were compensated by re-machining mold inserts. Finally, two molded V-beam structures were stacked up to construct a double V-beam device to produce bistable motion and testing was completed by measuring force-displacement relationships for every molded part.

2. Design of Double V-Beam Bistable Mechanism

2.1. Beam-Based Bistable Mechanisms (BM)

The V-beam BM was inspired by a stress-free, as-fabricated curved beam with an initial first buckling-mode shape. Figure 1a shows the single cosine curved beam that was made rectilinear with fixed ends. When a lateral force is applied, bistable motion is produced by post-buckling phenomena and the beam takes the shape of the first deflection mode. In this single-curved beam, however, the buckled beam in the first deflection mode moves back to its original shape when the lateral force is removed. Qiu et al. presented two criteria for bistable motion to occur, based on the buckling mode analysis of a curved beam: (1) the ratio of the curve height (apex height) to the beam thickness is large and (2) the asymmetric second mode must be constrained because the force with the second mode acts in the direction to original state of the beam [2]. To obtain stable bistability with a single-curved beam, compressive internal stress needs to be induced. Generally, two or several parallel beams connected at their centers are used to constrain the secondary mode and produce robust bistability, since adding axial loads would complicate a system, as shown in Figure 1b. The straight parallel V-beam structure in Figure 1c was manufactured using in-plane processes such as surface micromachining and WEDM.

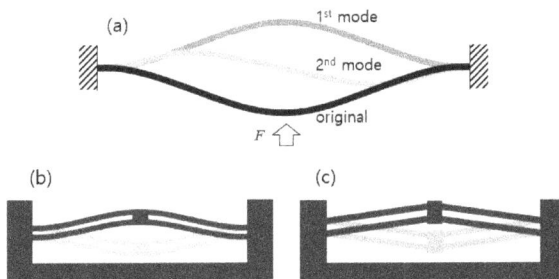

Figure 1. Compliant bistable mechanisms: (**a**) First and second post-buckling states of a pre-compressed beam; (**b**) curved parallel beams; (**c**) V-shaped parallel beams.

Fabrication of three-dimensional compliant members with force transfer and/or out-of-plane displacement is generally challenging, especially in miniaturized devices, leading to the design of complex 2.5D structures. Substrate out-of-plane BMs are very useful in various micro-systems. Beam buckling and bimorph effect of in-plane structure multi-materials have been used for out-of-plane motion, but in general, the creation of three-dimensional structure of at a micro level is quietly complex.

The V-beam bistable behavior can be expressed as the relationship between force and displacement, as shown in Figure 2. The first stable point is when the force does not have an effect, and as it increases to a maximum, it reaches the first critical-buckling force. After this point, we observe two different behaviors. In a bistable system, over the negative stiffness region, the beam reaches a snap-through point where the force starts to act opposite to the first stable point and quickly shifts to the second bistable point, passing a minimum value, the second critical-buckling force. At the second bistable point, the V-beam remains bent without the effect of an external force. In a non-bistable system, the beam never reaches the negative force, even as the displacement increases. The beam moves back to original position when the force is eliminated.

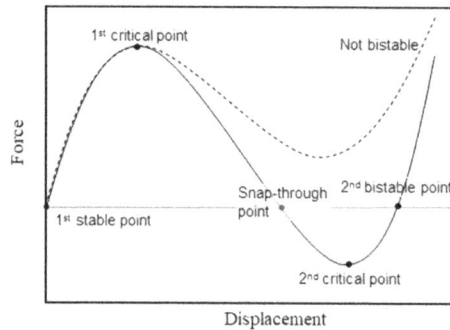

Figure 2. Force-displacement relation of a V-beam structure.

Force-displacement modelling and the resulting bistability predictability of pre-shaped beam structures were performed by the numerical solution for the differential equations of a post-buckling beam as well as the buckling mode analysis which can consider multiple (symmetric and asymmetric) buckling modes [2,12,13,15,29]. When the asymmetric mode can be constrained like in the double V-beam structure, the non-linear static FEA is also used for the force-displacement behavior [30–33].

2.2. Design Consideration for Injection Molding (IM) of Single V-Beam and Bistability

In the design stage, moldability and bistability should be considered simultaneously for successful fabrication. Essential features of compliant microsystem bodies are thin elastic beam segments. While beam slenderness is required, the attainable thickness-to-length ratio is limited because of its increased flow resistance of thermoplastic melt through cavities during a melt-filling stage. Consequently, the molded structure needs to be designed based on realistic data. In the author's previous studies, a cantilever 50 μm thick and 2 mm long was made using a two-stage mold containing a cavity of beams formed along the parting plane with thermoplastic polystyrene (PS) and polyoxymethylene (POM) resin, using a mold temperature controller [23]. A longer beam could be moldable, although severe flashes appeared unavoidably on its edge along the parting plane, and low strength caused beam deformation during ejection. Beam thickness was set to 150 μm, and other parameters such as length and V-beam inclined angles were determined from nonlinear FEA, which evaluates the given structure and POM material force–displacement relationship (elastic modulus: 2570 MPa, yield strength: 62 MPa).

As described in Section 2.1, double centrally clamped beams are required with a high apex height to bean thickness ratio to achieve the bistable snap-through behavior. However, direct injection molding of double V-beams aligned in out-of-plane direction is considerably challenging because undercuts appear between two beams, preventing their ejection from the mold. Therefore, we designed a single V-beam structure as a molded part first and analyzed the force-displacement relation using nonlinear structural static FEA (ANSYS workbench 19, ANSYS, Inc., Canonsburg, PA, USA) for two single V-beams structures that were stacked and bonded into double V-beam structures.

Figure 3a shows a sectional drawing of a designed bistable structure. Two symmetrically inclined beams of 5.2 mm long form a V-beam with a central shuttle. They connect to a rectangular main frame 1.5 mm thick through a 1-mm-long horizontal beam segment, including lower taper. The latter is for smooth thermoplastic melt flow from thick to thin and to lower pressure loss and residual stress after solidification. The beam width in the ground direction was designed with two values: 0.2 mm and 0.3 mm and was 0.15 mm thick. The whole BM consisted of three parallel V-beams having the same center shuttle, fixed to the main frame (18.8 mm × 11.5 mm × 1.5 mm), as shown in Figure 3b.

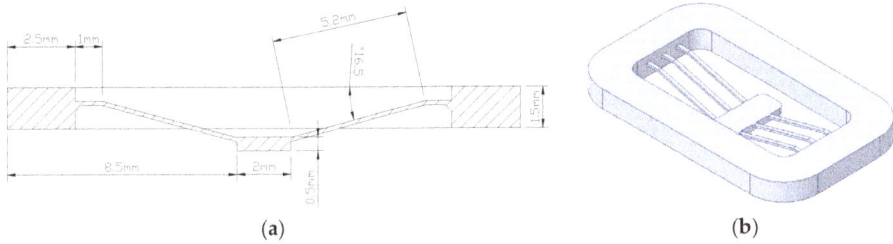

Figure 3. Design of single V-beam structures as mold part: (**a**) Cross section drawing; (**b**) 3D model.

Figure 4a presents a three-dimensional cross section of the two-stacked model shown in Figure 3b. A connecting insert is placed between two center shuttles and put in bonded contact with the faces of two center shuttles (the insert, the same POM as the structure here, is replaced with glue, as described later in Section 4.2). Two main frames were also put in boned contact with each other. A half-symmetry model of one double V-beam structure was built with the boundary conditions, as shown in Figure 4b. Hexahedral meshing was applied to the model and the total number of elements were 8190 for the 0.3 mm wide V-beam model and 7410 for the 0.2 mm model, respectively. The displacement of 2.8 mm was applied with 24 loading steps (0.02 mm successive increments to 0.2 mm, the 0.2 mm increments up to 2.8 mm). The calculated displacement as a function of applied force for the designed BM with three double V-beams is shown in Figure 4c. It is known that it has enough bistability.

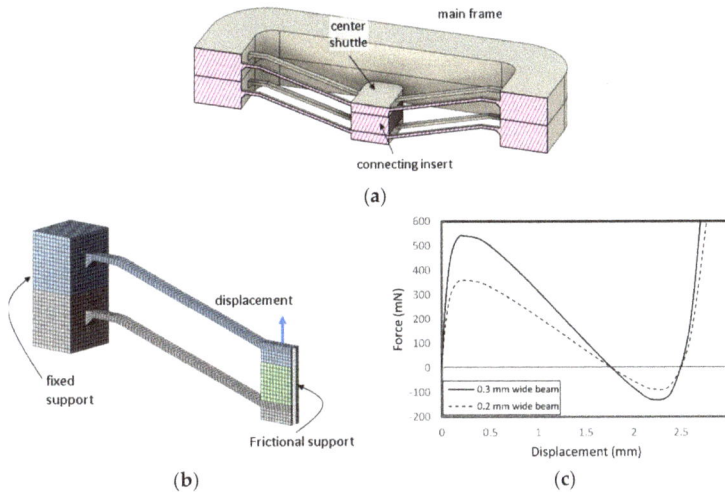

Figure 4. OBM based on V-beam structure stacking: (**a**) 3D sectional view; (**b**) FE model with boundary conditions; (**c**) simulated static force-displacement curves.

3. Fabrication

3.1. Layout and Tooling

A melt delivery system was designed using a symmetric two-cavity layout, as shown in Figure 5, considering the filling pattern of thermoplastic melt in mold cavities. Flow, cooling and solidification different in melt delivery can cause non-uniform material properties of the molded part. Gates (width 2 mm, thickness 0.8 mm, land 1 mm) were placed at the center of the long side of the main frame so that the beams on both sides of the center shuttle were subjected to the same conditions, as weld lines are formed inside where the fronts of two melt flows meet.

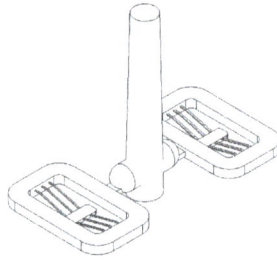

Figure 5. Melt delivery system.

The runner diameter is 4 mm and its length (from the sprue center to the gate) is 5.8 mm. An alloy of beryllium copper was used for the unibody insert, containing the melt delivery system and the part cavities, and fabricated using a high-speed milling machine. The parting line was created along the top edge of the V-beam and the cavities formed in a movable sided mold insert that is shown in Figure 6. Upper and lower cavities were for a 0.3 mm and a 0.2 mm wide beam, respectively.

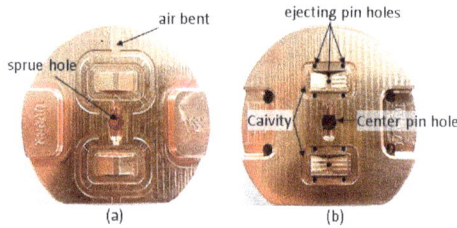

Figure 6. Machined mold insert: (**a**) fixed side; (**b**) movable side.

A 3-D digital microscope (Keyence VHX-900F, Keyence Corporation, Osaka, Japan) was used to measure the micro-cavities dimensions and the results are shown in Figure 7. The cavity sections were close to trapezoidal form and the corners were round at the bottom edge, presumably caused by tool deflection [28,29]. The beam cavities widths at the midpoints of the thickness were 330.5 μm and 238.1 μm respectively, and their thicknesses about 144 μm and 142 μm. Even though dimensional errors in beam cavities are directly associated to the molded beams and their stiffness, their effect is negligible as the moment of inertia for each beam changes less than 3%. The surface roughness on the cavity bottom is Rms 0.25 μm. The ejector pins contact four points on the rectangular frame and one pin was pushed up the center shuttle.

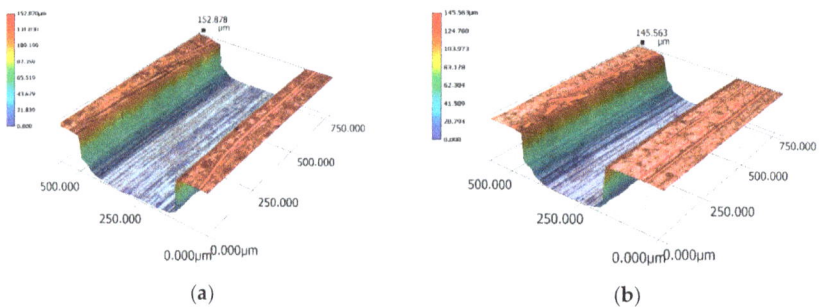

Figure 7. Three dimensional digital microscope images of machined cavities for 0.3 mm (**a**) and 0.2 mm (**b**) wide beam.

3.2. Material and Injection Molding

The polymer material used in IM is a POM (Lucel N109, melt flow index: 9 g/10 min, supplied by LG Chemistry, Seoul, Korea), the semi-crystalline engineering plastic. POM has been widely used in MM research due to its processability, such as low viscosity and thermal stability. POM is also known for its high-tensile strength, rigidity, high fatigue resistance, natural lubricity, and environmental stability. An IM machine with a vertical single plunger and a diameter of 16 mm (LS-30, Canon Electronics, Tokyo, Japan) was used. Its maximum injection speed is 75 mm/s and clamp force is 29 kN.

In order to fill the cavities completely through the melt delivery system in Figure 4, high settings for injection speed and holding pressure are required. A fast mold temperature control is also indispensable, in which heater rods and chilled air are applied into the mold insert of Figure 5 [23,26].

The process parameter values applied for a successfully molded BM are shown in Table 1.

Table 1. Parameter settings for complete filling.

Process Parameters	Value
Injection speed (mm/s)	75
Injection pressure (MPa)	75
Pressure holding time (s)	4
Melt temperature (°C)	220
Mold temperature (°C)	130 (filling), 100 (ejecting)

Figure 8 shows the short shots obtained by increasing the injection pressure from 20 MPa to 65 MPa, filling the micro-cavity by the stagnant pressure in the main cavity. The left side of the central sprue contains the micro-cavities of the 0.3 mm wide beam and the right side contains those of the 0.2 mm wide beam. As shown in Figure 8a, filling begins in all micro-cavities on both sides and its length increases symmetrically but slightly faster in the left side due to its higher width, described in Figure 8b–d. Figure 8e shows the two-melt fronts meeting at the central shuttle in the left-side cavity, but the weld lines are out of center in the right one, which might have been caused by asymmetrical-cavity dimensional error. Such unbalanced filling gives rise to asymmetrical deformation of the molded V-beam when a force acts on the center shuttle after.

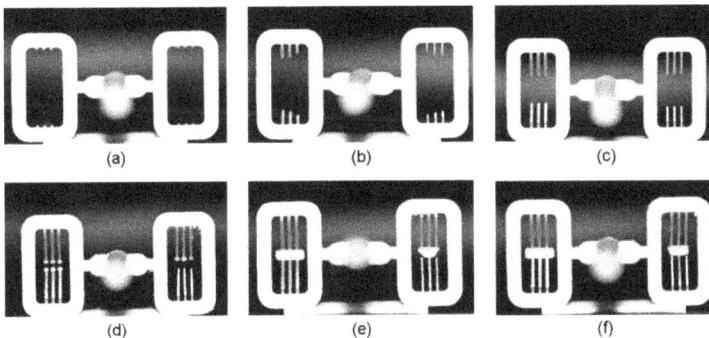

Figure 8. Short shots obtained during melt filling stage, by increasing injection pressure with other factors fixed (Injection pressure of (**a**) 20, (**b**) 25, (**c**) 30, (**d**) 45, (**e**) 55, and (**f**) 65 MPa).

Flash is a common defect in IM caused not only by an excessive cavity pressure over the nominal clamping force, but plastic melt high flowability due to high mold or melt temperature. Especially when molding a micro-beam, flash can easily occur under high settings of mold/melt temperature and injection speed [24,27]. In our experiment, severe flash was encountered when the mold temperature controller increased to 140 °C at filling, as shown in Figure 9. It formed even at 130 °C but this amount of flash did not adversely affect the deformation. To obtain a part with no defects, successful molding condition should be set carefully because the range is very narrow between short shot and flash occurrence.

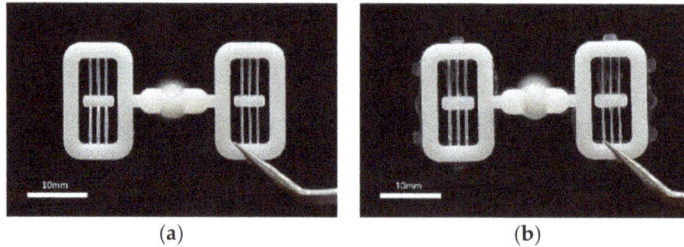

(a) (b)

Figure 9. Molded V-beam structures in good quality (**a**) and with flashes (**b**).

3.3. Shrinkage and Shape Error Compensation

The elimination of thermal shrinkage of molded parts during cooling and ejection is likely a challenge even when melt fills completely the micro-cavities. Thermal shrinkage causes V-beam geometric errors from the original design, which needs to be minimized as to not affect the bistability. As described in Figure 10a, it was assumed that shrinkage occurs towards the center of the part, and main frame and beams longitudinal shrinkages give rise to vertical position changes of the center shuttle as well as of the inclined beam angles (Δh and $\Delta\theta = \theta - \theta'$). It is also assumed that the shrinkage ratios are different between the beam and the main frame because a thinner beam cools much faster.

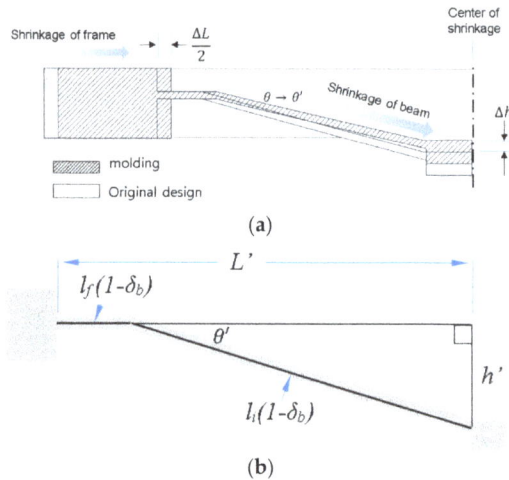

(a)

(b)

Figure 10. Model of shrinkage for mold compensation: (**a**) shrinkage and change of geometry (half of molded part); (**b**) notations of dimension for shrunk structure.

The length of the main frame was measured with a profile projector (PJ-A3000, Mitutoyo, Kawasaki, Japan) taking the average of six molded samples (The distance is measured between the edges of the fixed end of a beam) and used to determine the shrinkage ratio along the same length in the mold insert. Instead of measuring the V-beam length directly due to it difficulty, a height gauge was used to measure the vertical deviation of the center shuttle. Taking the shrinkage ratio in the longitudinal direction of the micro-beam δb, the right-angled triangle relation can be applied to find δb, as shown in Figure 10b. The original lengths of the horizontal and inclined sections of the beam are lf and ls. The horizontally shrunken length of the main frame in the region of the beam is L', and h' is the vertical position of the center shuttle, which can be determined by measuring the gap between the two horizontal planes formed by the main frame and the center shuttle. The inclined angle of the beam in the molded part is θ'.

Using the Pythagorean Theorem, beam shrinkage ratios are shown in Table 2. The ratio for the beam was 1.08%, smaller than that of the main frame at 1.78%. The general shrinkage ratio from the supplier is 1.8%–2.1%, close to the main frame value. We modified the dimensions to reduce a large error in h, θ, and L. Those for the compensated mold and those resulting under the same conditions used for the first molding experiment are shown in Table 2. The shrinkage is similar in both the main frame and the beam. The error values compared to the original design were clearly reduced.

Table 2. Dimensions of mold and molded parts of the original and the compensated mold, whose errors were obtained by comparing original design dimensions.

Design	Geometric Variable	Mold	Part	Shrinkage (%)	Error (%)
Original	L (mm)	5.83	5.726	1.78	−1.78
	l_f (mm)	1.0	0.99	1.08	−1
	l_i (mm)	5.0	4.95		−1
	h (mm)	1.294	1.175	–	−9.1
	θ (°)	15.0	13.74	–	−8.4
Compensation	L (mm)	5.99	5.869	2.02	0.6
	l_f (mm)	1	0.99	1.01	−1
	l_i (mm)	5.2	5.15		3
	h (mm)	1.45	1.34	–	3.6
	θ (°)	16.4	15.08	–	0.6

4. Experiment

4.1. Setup for Measurement of Force-Displacement Behavior

To obtain the relationship between the driving force and the center-shuttle vertical displacement in molded V-beam structures, a digital force gauge (ZTA-5N, Imada, force resolution 1 mN, maximum force capacity 5 N) was fixed on a linear stage so its probe can press the shuttle central point, as shown in Figure 11. Both molded main-frame sides were attached to a block with a XY-stage-mounted groove. After touching the center of the shuttle, the probe moved down further 2.8 mm while deforming the V-beam. Force values of 10 as-molded samples were measured at intervals of 1 mm from the point of contact.

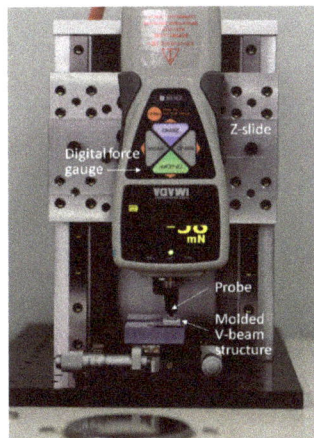

Figure 11. Setup of the force-displacement measurement experiment.

4.2. Force–Displacement Relationship of Molded V-beam Structures

Figure 12 shows the measured force-displacement curves for the ten as-molded single V-beam structures. In the curves, the force increases with the displacement in the early region and after the peak point, the measured forces show negative stiffness, though they do not become negative, failing to produce bistable motion as previously discussed in Section 2.1. This is because the unconstrained, S-shaped second mode was produced, thus preventing snap-through bistable motion.

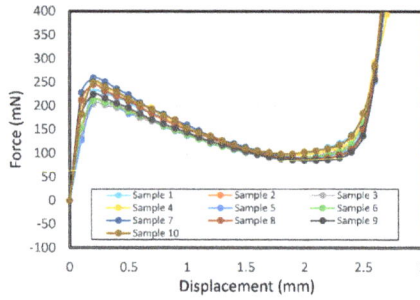

Figure 12. Experimental force–displacement relationship for ten molded samples of single V-beam structure.

We constructed a double V-beam structure OBM by stacking up two molded V-beams structures using hot melt adhesive to glue the top and bottom surfaces of the main frames and the center shuttle together, as shown in Figure 13.

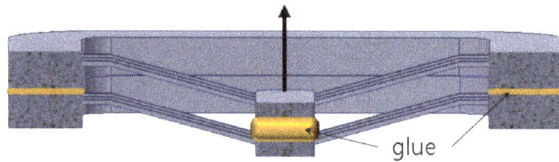

Figure 13. Double V-beam structure made by stacking up two molded single V-beam parts.

Figure 14 shows a graph of its experimental behavior from three fabricated samples. The peak force doubles with respect to the displacement of a single V-beam structure. All assembled-double-V-beam structures show clear snap-through behavior and bistability.

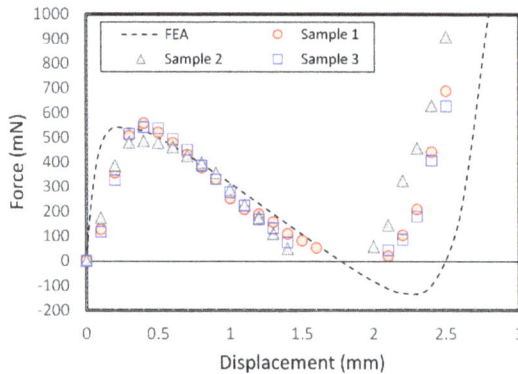

Figure 14. Experimental force-displacement relationship for three assembled double V-beam samples.

Negative stiffness leads to zero at displacements of 1.5–2 mm, caused by quick snap-through action. The negative force data did not appear on the graph because the shuttle was detached from the probe when the snapping action occurred. The experimental results show some discrepancies with the numerical solution, especially in the early displacement and after the snap-through point, which may be due to soft adhesive layer connecting two parts and relative position errors. If assembling the two structures was more improved, the behavior would be more predictable. The bistability of the assembled double V-beam structure is shown in Figure 15, in which the V-beams are at two stable points.

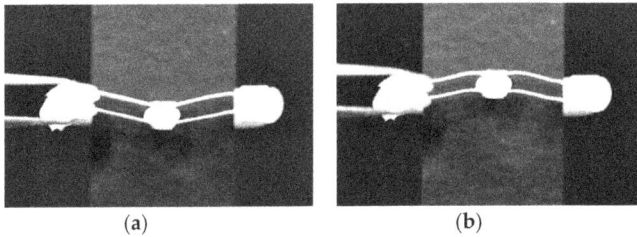

(a)　　　　　　　　　　　　　　(b)

Figure 15. Photograph of the structure (front main frame segment is cut off and a black paper strip covers the rear frame to make the micro-beams clearly visible) in (**a**) the original position and (**b**) the second stable position after snap-through.

5. Conclusions

Our study highlights that miniaturized plastic BMs based on compliant V-beam-like structures can be fabricated with good quality by MM technology and stacking up two V-beam structures. In the MM process, thermoplastic melt fills completely the micro-beam cavities of the high length-to-thickness ratio over 40 using high-temperature conditions with design for injection-molding. If bistable motion can be extended to multiple dimensions with balanced kinetic characteristics through mechanism design research and/or automatic motion can be produced by smart or stimuli responsive designs using like shape memory polymers, the high productivity and low cost characteristics of the injection-molding process could eventually lead to its application in various microsystems [34–37].

Recently, architected materials, defined as materials whose effective properties result from their ordered microarchitecture rather than their intrinsic material properties have been attracting considerable interest [38–43]. Fast advances in additive manufacturing technologies have enabled precise fabrication of new types of complex architected materials. Periodic BMs with symmetric slender beams can be applied to reversible energy-absorbing and tunable morphological changing architected materials, opening a range of new functionalities. We hope that a bottom-up assembly process can be used as the fabrication technique for three-dimensional functional materials along with the advanced additive manufacturing techniques. Furthermore, BMs fabricated by MM have enough potential to be used as a unit cell if a modular design plan and the optimized mechanical response are provided, together with the proper part design of the assembly [44,45].

Author Contributions: Conceptualization, W.-B.K. and S.-Y.H.; methodology, W.-B.K. and S.-Y.H.; software, S.-Y.H.; validation, W.-B.K.; formal analysis, W.-B.K. and S.-Y.H.; investigation, W.-B.K. and S.-Y.H.; writing—original draft preparation, S.-Y.H.; writing—review and editing, W.-B.K.; visualization, W.-B.K. and S.-Y.H.; supervision, W.-B.K.; project administration, W.-B.K.; funding acquisition, W.-B.K. All authors have read and agreed to the published version of the manuscript.

Funding: This work was supported by the National Research Foundation of Korea (NRF) grant funded by the Korea government (MEST) NRF-2018R1D1A1B07050525.

Conflicts of Interest: The authors declare no conflict of interest.

References

1. Baker, M.S.; Howell, L.L. On-chip actuation of an in-plane compliant bistable micromechanism. *J. Microelectromech. Syst.* **2002**, *11*, 566–573. [CrossRef]
2. Qiu, J.; Lang, J.H.; Slocum, A.H. A curved-beam bistable mechanism. *J. Microelectromech. Syst.* **2004**, *13*, 137–146. [CrossRef]
3. Casals-Terre, J.; Fargas-Marques, A.; Shkel, A.M. Snap-action bistable micromechanisms actuated by nonlinear resonance. *J. Microelectromech. Syst.* **2008**, *17*, 1082–1093. [CrossRef]
4. Hansen, B.J.; Carron, C.J.; Jensen, B.D.; Hawkins, A.R.; Schultz, S.M. Plastic latching accelerometer based on bistable compliant mechanisms. *Smart Mater. Struct.* **2007**, *16*, 1967–1972. [CrossRef]
5. Megnin, C.; Barth, J.; Kohl, M. A bistable SMA microvalve for 3/2-way control. *Sens. Actuators Phys.* **2012**, *188*, 285–291. [CrossRef]
6. Hu, N.; Burgueño, R. Buckling-induced smart applications: Recent advances and trends. *Smart Mater. Struct.* **2015**, *24*, 063001. [CrossRef]
7. Harne, R.L.; Wang, K.W. A review of the recent research on vibration energy harvesting via bistable systems. *Smart Mater. Struct.* **2013**, *22*, 023001. [CrossRef]
8. Nguyen, S.D.; Halvorsen, E.; Paprotny, I. Bistable springs for wideband microelectromechanical energy harvesters. *Appl. Phys. Lett.* **2013**, *102*, 023904. [CrossRef]
9. Saif, M.T.A. On a tunable bistable MEMS-theory and experiment. *J. Microelectromech. Syst.* **2000**, *9*, 157–170. [CrossRef]
10. Ongaro, F.; Jin, Q.; Siciliani de Cumis, U.; Ghosh, A.; Denasi, A.; Gracias, D.H.; Misra, S. Force characterization and analysis of thin film actuators for untethered microdevices. *AIP Adv.* **2019**, *9*, 055011. [CrossRef]
11. Zirbel, S.A.; Tolman, K.A.; Trease, B.P.; Howell, L.L. Bistable mechanisms for space applications. *PLoS ONE* **2016**, *11*, e0168218. [CrossRef] [PubMed]
12. Wu, C.-C.; Lin, M.-J.; Chen, R. The derivation of a bistable criterion for double V-beam mechanisms. *J. Micromech. Microeng.* **2013**, *23*, 115005. [CrossRef]
13. Pustan, M.; Chiorean, R.; Birleanu, C.; Dudescu, C.; Muller, R.; Baracu, A.; Voicu, R. Reliability design of thermally actuated MEMS switches based on V-shape beams. *Microsyst. Technol.* **2017**, *23*, 3863–3871. [CrossRef]
14. Hayes, G.R.; Frecker, M.I.; Adair, J.H. Fabrication of compliant mechanisms on the mesoscale. *Mech. Sci.* **2011**, *2*, 129–137. [CrossRef]
15. Todd, B.; Jensen, B.D.; Schultz, S.M.; Hawkins, A.R. Design and testing of a thin-flexure bistable mechanism suitable for stamping from metal sheets. *J. Mech. Des.* **2010**, *132*, 071011. [CrossRef]
16. Young, R.J.; Lovell, P.A. *Introduction to Polymers*; CRC Press: Boca Raton, FI, USA, 2011.
17. Giboz, J.; Copponnex, T.; Mélé, P. Microinjection molding of thermoplastic polymers: Morphological comparison with conventional injection molding. *J. Micromech. Microeng.* **2009**, *19*, 025023. [CrossRef]
18. Yang, C.; Yin, X.-H.; Cheng, G.-M. Microinjection molding of microsystem components: New aspects in improving performance. *J. Micromech. Microeng.* **2013**, *23*, 093001. [CrossRef]
19. Lucchetta, G.; Masato, D.; Sorgato, M.; Crema, L.; Savio, E. Effects of different mould coatings on polymer filling flow in thin-wall injection moulding. *CIRP Ann.* **2016**, *65*, 537–540. [CrossRef]
20. Griffiths, C.A.; Rees, A.; Kerton, R.M.; Fonseca, O.V. Temperature effects on DLC coated micro moulds. *Surf. Coat. Technol.* **2016**, *307*, 28–37. [CrossRef]
21. Xu, G.; Yu, L.; Lee, L.J.; Koelling, K.W. Experimental and numerical studies of injection molding with microfeatures. *Polym. Eng. Sci.* **2005**, *45*, 866–875. [CrossRef]
22. Sha, B.; Dimov, S.; Griffiths, C.; Packianather, M.S. Investigation of micro-injection moulding: Factors affecting the replication quality. *J. Mater. Process. Technol.* **2007**, *183*, 284–296. [CrossRef]
23. Kim, W.-W.; Gang, M.G.; Min, B.-K.; Kim, W.-B. Experimental and numerical investigations of cavity filling process in injection moulding for microcantilever structures. *Int. J. Adv. Manuf. Technol.* **2014**, *75*, 293–304. [CrossRef]
24. Eladl, A.; Mostafa, R.; Islam, A.; Loaldi, D.; Soltan, H.; Hansen, H.N.; Tosello, G. Effect of process parameters on flow length and flash formation in injection moulding of high aspect ratio polymeric micro features. *Micromachines* **2018**, *9*, 58. [CrossRef]

Micromachines **2020**, *11*, 155

25. Mélé, P.; Giboz, J. Micro-injection molding of thermoplastic polymers: Proposal of a constitutive law as function of the aspect ratios. *J. Appl. Polym. Sci.* **2018**, *135*, 45719. [CrossRef]

26. Han, S.-Y.; Kim, W.-B. Microinjection moulding of miniaturised polymeric ortho-planar springs. *Microsyst. Technol.* **2016**, *22*, 1991–1999. [CrossRef]

27. McFarland, A.W.; Poggi, M.A.; Bottomley, L.A.; Colton, J.S. Injection moulding of high aspect ratio micron-scale thickness polymeric microcantilevers. *Nanotechnology* **2004**, *15*, 1628–1632. [CrossRef]

28. Urwyler, P.; Schift, H.; Gobrecht, J.; Häfeli, O.; Altana, M.; Battiston, F.; Müller, B. Surface patterned polymer micro-cantilever arrays for sensing. *Sens. Actuators Phys.* **2011**, *172*, 2–8. [CrossRef]

29. Zhou, W.; Yu, H.; Chen, L.; Chen, Y.; Peng, B.; Peng, P. Stability analysis of two coupled pre-shaped beams in parallel. *Proc. Inst. Mech. Eng. Part C J. Mech. Eng. Sci.* **2018**, *232*, 2482–2489. [CrossRef]

30. Wilcox, D.L.; Howell, L.L. Fully compliant tensural bistable micromechanisms (FTBM). *J. Microelectromech. Syst.* **2005**, *14*, 1223–1235. [CrossRef]

31. Xu, Q. Design of a large-stroke bistable mechanism for the application in constant-force micropositioning stage. *J. Mech. Robot.* **2017**, *9*, 011006. [CrossRef]

32. Masters, N.D.; Howell, L.L. A self-retracting fully compliant bistable micromechanism. *J. Microelectromech. Syst.* **2003**, *12*, 273–280. [CrossRef]

33. Yan, W.; Yu, Y.; Mehta, A. Analytical modeling for rapid design of bistable buckled beams. *Theor. Appl. Mech. Lett.* **2019**, *9*, 264–272. [CrossRef]

34. Ghosh, A.; Yoon, C.; Ongaro, F.; Scheggi, S.; Selaru, F.M.; Misra, S.; Gracias, D.H. Stimuli-responsive soft untethered grippers for drug delivery and robotic Ssurgery. *Front. Mech. Eng.* **2017**, *3*, 7. [CrossRef] [PubMed]

35. Tolou, N.; Henneken, V.A.; Herder, J.L. Statically balanced compliant micro mechanisms (SB-MEMS): Concepts and simulation. In Proceedings of the ASME 2010 International Design Engineering Technical Conferences and Computers and Information in Engineering Conference, Montreal, QC, Canada, 15–18 August 2010; pp. 447–454.

36. Synthesis of Mechanisms with Prescribed Elastic Load-Displacement Characteristics | TU Delft Repositories. Available online: https://repository.tudelft.nl/islandora/object/uuid:d518b379-462a-448f-83ef-5ba0e761c578 (accessed on 23 January 2020).

37. Lamers, A.J.; Gallego Sánchez, J.A.; Herder, J.L. Design of a statically balanced fully compliant grasper. *Mech. Mach. Theory* **2015**, *92*, 230–239. [CrossRef]

38. Shan, S.; Kang, S.H.; Raney, J.R.; Wang, P.; Fang, L.; Candido, F.; Lewis, J.A.; Bertoldi, K. Multistable architected materials for trapping elastic strain energy. *Adv. Mater.* **2015**, *27*, 4296–4301. [CrossRef]

39. Che, K.; Yuan, C.; Qi, H.J.; Meaud, J. Viscoelastic multistable architected materials with temperature-dependent snapping sequence. *Soft Matter* **2018**, *14*, 2492–2499. [CrossRef]

40. Yang, H.; Ma, L. Multi-stable mechanical metamaterials by elastic buckling instability. *J. Mater. Sci.* **2019**, *54*, 3509–3526. [CrossRef]

41. Haghpanah, B.; Salari-Sharif, L.; Pourrajab, P.; Hopkins, J.; Valdevit, L. Multistable shape-reconfigurable architected materials. *Adv. Mater.* **2016**, *28*, 7915–7920. [CrossRef]

42. Chen, T.; Mueller, J.; Shea, K. Integrated design and simulation of tunable, multi-state structures fabricated monolithically with multi-material 3D printing. *Sci. Rep.* **2017**, *7*, 1–8. [CrossRef]

43. Wu, X.; Su, Y.; Shi, J. Perspective of additive manufacturing for metamaterials development. *Smart Mater. Struct.* **2019**, *28*, 093001. [CrossRef]

44. Cao, L.; Dolovich, A.T.; Schwab, A.L.; Herder, J.L.; Zhang, W. (Chris). Toward a unified design approach for both compliant mechanisms and rigid-body mechanisms: Module optimization. *J. Mech. Des.* **2015**, *137*, 122301. [CrossRef]

45. Cao, L.; Dolovich, A.T.; Chen, A.; Zhang, W. (Chris). Topology optimization of efficient and strong hybrid compliant mechanisms using a mixed mesh of beams and flexure hinges with strength control. *Mech. Mach. Theory* **2018**, *121*, 213–227. [CrossRef]

micromachines

MDPI

Review

State-of-the-Art and Perspectives on Silicon Waveguide Crossings: A Review

Sailong Wu, Xin Mu, Lirong Cheng, Simei Mao and H.Y. Fu *

Tsinghua-Berkeley Shenzhen Institute, Tsinghua University, Shenzhen 518055, China;
wusl17@mails.tsinghua.edu.cn (S.W.); mux17@mails.tsinghua.edu.cn (X.M.); clr18@mails.tsinghua.edu.cn (L.C.);
maosm19@mails.tsinghua.edu.cn (S.M.)
* Correspondence: hyfu@sz.tsinghua.edu.cn; Tel.: +86-755-3688-1498

Received: 10 February 2020; Accepted: 19 March 2020; Published: 20 March 2020

Abstract: In the past few decades, silicon photonics has witnessed a ramp-up of investment in both research and industry. As a basic building block, silicon waveguide crossing is inevitable for dense silicon photonic integrated circuits and efficient crossing designs will greatly improve the performance of photonic devices with multiple crossings. In this paper, we focus on the state-of-the-art and perspectives on silicon waveguide crossings. It reviews several classical structures in silicon waveguide crossing design, such as shaped taper, multimode interference, subwavelength grating, holey subwavelength grating and vertical directional coupler by forward or inverse design method. In addition, we introduce some emerging research directions in crossing design including polarization-division-multiplexing and mode-division-multiplexing technologies.

Keywords: silicon photonics; integrated optics; waveguide crossing design; multimode interference (MMI); sub-wavelength grating (SWG); multiplexing technology

---

1. Introduction

Towards the increasing demands of higher data rates, the issues of electric signal attenuation and power dissipation rise dramatically due to the intrinsic limitation of the parasitic effects in current metallic interconnection [1]. Moore's law, the principle that has powered the information-technology revolution since the 1960s, is approaching its growth limit [2]. Fortunately, photons that have zero rest mass and zero charge, and can travel at the speed of light without interfering with electromagnetic field, is nearly 1000 times faster than electrons. The photonics integrated circuits (PICs) have greater advantages over the traditional integrated circuits (ICs) including higher data rate, larger bandwidth and lower power consumption. Hybrid technologies utilizing both PICs and ICs are advancing, enabling next generation science and technology for information society, pushing the boundaries of what is possible for telecommunications, computing, defense and consumer technology [3–5].

Silicon is a semiconductor that can both conduct electrons as a conductor or function as an insulator by controlling the charge and number of activated carriers in the doping processes, which makes silicon an ideal material to become the basis of memory chips, powering various devices from portable calculators to supercomputers. On the other hand, silicon is the second most abundant element on the earth. Besides, high-yield, large-scale silicon electronic devices can be manufactured with the current mature complementary metal-oxide semiconductor (CMOS) techniques. In terms of optical property, silicon is transparent for wavelengths from 1.1 to 8 μm, which covers near-infrared (NIR) and parts of mid-infrared (MIR) region [6]. The NIR covering the entire original-band (O-band) and conventional-band (C-band) with extremely low attenuation has been well studied for current fiber-optic communications. The MIR silicon photonics, which includes both atmospheric window (3–5 μm) and absorption bands (2.6–2.9 μm and >3.6 μm) of most chemical and biological molecules, as well as fingerprint region has recently been studied for optical interconnect and spectroscopic sensing, respectively [7,8]. As a result, silicon is regarded as an excellent candidate for the marriage of PICs and ICs on the same CMOS platform. The term "silicon photonics" refers to the applications of photonic systems which use silicon as an optical or sensing medium [9–11].

Silicon photonics, named by Soref, can date back to mid-1980s and began commercialization by Bookham Technology Ltd. in 1989 [12,13]. An explosive developments of silicon photonics have been witnessed in recent decades, revolutionizing a number of application areas, for example, data centers, high-performance computing and sensing [14,15]. The silicon-on-insulator (SOI) platform is promising due to its high refractive index contrast and the compatibility with commercial CMOS technology. In order to design dense and fully functional photonic components on a SOI platform, silicon waveguide crossing is critical and inevitable when the system is becoming more and more complexed, while, on the other hand, a compact device footprint is increasingly in demand. For electrical circuits, most of the printed circuit boards (PCBs) have four to eight layers and supercomputers typically contain boards with more than sixteen layers to improve the computation functionalities [16]. Controlling the PCB layers is a very flexible and mature technique for balancing the device performance and power consumption. However, efficient optical vias for multiple layers are very difficult to implement in the high-index contrast SOI platform. Therefore, this method cannot be employed in the silicon photonic circuits due to the limitations of optical mode coupling and fabrication cost [17]. For a typical direct silicon waveguide crossing design, the insertion loss is around 1.4 dB and the crosstalk is −9.2 dB [18], which means that the optical power nearly shrinks into a half after passing through only two cascaded silicon waveguide crossings. This kind of inefficient waveguide crossing design will greatly aggravate the performance of the advanced PICs devices with many cascaded waveguide crossings involved, such as optical routers [19,20]. In this review paper, we discuss and summarize different kinds of silicon waveguide crossing designs (e.g., shaped taper, multimode interference, subwavelength grating and vertical directional coupler structure). In addition, we introduce several recent hot research topics for the waveguide crossing design (e.g., polarization-multiplexing technology and mode-multiplexing technology).

2. The Key Technologies of Silicon Waveguide Crossing

Silicon waveguide usually consists of a 2 μm silica lower cladding, 220 nm silicon core and silica upper cladding on the 200 mm (8 inch) wafers, as shown in Figure 1a. It is fabricated by the deep ultraviolet (DUV) lithography and this technique can provide fast and reliable patterns. The refractive indexes of silicon and silica are 3.45 and 1.44, respectively, at 1550 nm, and the high refractive index contrast between the core and cladding results in strong optical confinement and ultra-small bending radius for fundamental modes [21]. The dimension of silicon core fulfills the optical single-mode condition and the experimental propagation losses of the fundamental mode are 2.4 ± 0.2 and 0.59 ± 0.32 dB/cm for transverse-electric (TE) and transverse-magnetic (TM) modes, respectively [22,23]. For the low refractive index contrast structure, such as optical fiber, direct waveguide crossings are just a minor perturbation of the straight waveguide [24]. However, side effects of the direct waveguide crossings cannot be ignored in the SOI platform for the beam is dramatically diffracted in the silicon intersection region, as shown in Figure 1b.

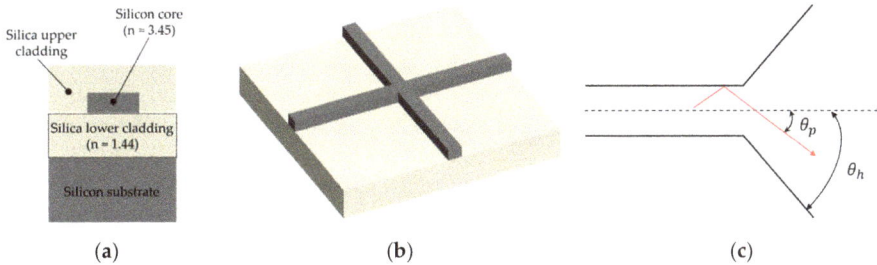

Figure 1. (**a**) Cross section of a typical silicon photonic wire waveguide. The grey structure represents the material silicon and the tawny one represents the material silica. The following schematic figure obeys the same rule. (**b**) Scheme of the direct silicon waveguide crossing on silicon-on-insulator (SOI) platform. (**c**) The ray model of the fundamental mode in the waveguide. θ_p is the light projection angle and θ_h is the horn angle of the optical waveguide.

These phenomena can be approximately explained by the mode–conversion relationship between ray angle of the light and horn angle of the waveguide [25], as shown in Figure 1c. The silicon waveguide allows a total internal reflection of over 60° incidence angle due to the large refractive index contrast, while the horn angle jumps abruptly to 90° at intersection region for the direct waveguide crossing design. Huge mismatch between the projection angle and horn angle cannot maintain a smooth power transition. The large portions of light are excited into leaky higher-order modes and propagate into the vertical waveguides at the core region, resulting in the side effects of crosstalk and large insertion loss. From another point of view, the light is kept well confined in the silicon waveguide channel due to high refractive index difference before entering the "dangerous" core section. Once entering the intersection region, the horizontal silica cladding suddenly disappears and the optical power begins to scatter in all directions, resulting in severe effects of insertion loss, crosstalk and back-reflection. This section is organized to introduce several silicon waveguide crossing designs and explains their principles for improving the crossing performance.

2.1. Shaped Taper Method

Shaped taper waveguide crossing is to tailor the width of the silicon channel towards the center region, which decreases the diffraction effects by avoiding the shape mutation between the channel waveguide and crossing region. Shaped taper design includes different mathematic types, such as linear taper, parabolic taper, exponential taper and Gaussian taper [26]. With the shaped taper structure, the guided modes are expanded and the wide-angle spatial components are reduced, since the widths of waveguides are smoothly getting larger [27]. The diffraction side effects can be greatly controlled when the optical modes have fewer wide-angle spatial components.

The elliptical taper profile is used to get a narrow angular spectrum of the expanded mode, as shown in Figure 2a. The insertion loss and crosstalk of the device are <0.1 and <−30 dB with a footprint of $7.2 \times 1.5\ \mu m^2$, respectively [18]. For the nonadiabatic taper, the higher-order leaky optical modes will be excited and a sizable fraction of the power will be radiated away in the core region. The adiabatic taper can smoothly expand the guided modes, while the large taper footprint is not preferred for highly integrated PICs. As a result, a trade-off between its optical performance and footprint needs to be carefully considered. The double etching scheme consists of the high-contrast photonic wires in the upper level and the shadow-etched parabolic tapers in the lower level, as shown in Figure 2b. The confinement of the high refractive index contrast is maintained in the upper part and the lower parabolic expanded tapers play a role to adjust the optical phase fronts. The double DUV etching scheme waveguide crossing can greatly shrink the footprint to $6 \times 6\ \mu m^2$ and the insertion loss and crosstalk are 0.16 and −40 dB, respectively [28]. However, the additional DUV stepper lithography increases the fabrication complexity and cost of the dual-etching device are much higher than that of the conventional single DUV etching process, which are the main limitations of this kind of waveguide crossing design. Different from the assumed mathematical taper profile with only a few parameters to be determined, a numerical optimization method has more degrees of freedom by separating the entire taper into many spaced segments with different widths, as shown in Figure 2c. The shaped taper can be formed by connecting all the segments with the spline function and the optimization process is implemented to minimize the insertion loss and crosstalk by tuning the widths of all the segments with the help of advanced algorithms. The insertion loss and crosstalk are <0.2 and <−40 dB, respectively. The footprint is 6 × 6 µm^2 for the silicon waveguide crossing designed by genetic algorithm (GA) [29]. For particle swarm optimization algorithm (PSO) assisted silicon waveguide crossing design, the insertion loss and crosstalk are −0.0278 ± 0.0092 and <−37 dB, respectively. The device has a footprint of 9 × 9 µm^2 [30]. However, DUV steppers are not designed for high resolution purposes and the performance begins to degrade for small designs. The small numerical optimized shaped taper can be fabricated with single etch process, while the fabrication tolerance is small since the performance is sensitive to the device geometry. The design performance varies greatly in terms of the fabrication errors in DUV technology and this is the main drawback for this crossing design. For the conventional silicon waveguide crossing, the crossing angle is 90° and the device is consisted of two perpendicular arms. Researchers find that the crosstalk can be improved by more than 10 dB without degrading transmission losses where the crossing angle is set to be 60° for the direct waveguide crossing. For the double-etched waveguide crossing, the crosstalk can be reduced by 3.7 dB without degrading transmission losses by titling waveguide crossing angle to be 60° [31], as shown in Figure 2d. The offset crossing with a small angle of 20° is also proposed and proves to be beneficial for further reducing the crosstalk [32].

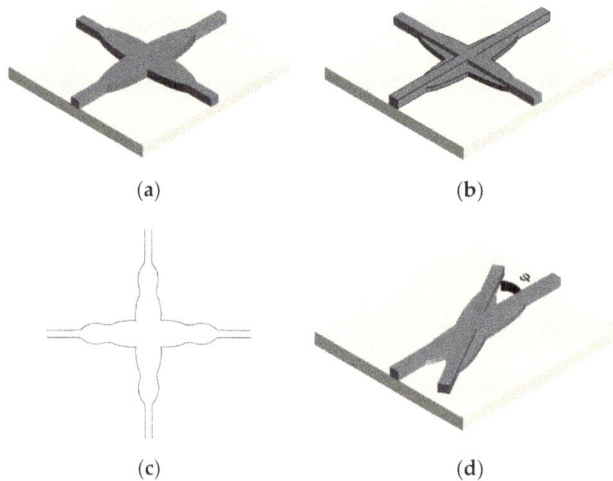

(a)   (b)

(c)   (d)

Figure 2. Schematics of (**a**) the shaped taper waveguide crossing and (**b**) the double-etched shaped taper silicon waveguide crossing; (**c**) the top view of the shaped waveguide crossing designed by the genetic algorithm (GA); (**d**) Schematic of the titled double-etched silicon waveguide crossing, φ represents the crossing angle.

2.2. Multimode Interference Method

The structure of the multimode interference (MMI) device is such that single-mode waveguide connects to both sides of a multimode waveguide, and there exists the self-imaging phenomena that an input signal pattern is replicated, at periodic intervals, once or multiple times along the direction of propagation along the waveguide [33]. If the light is launched at the center position of the waveguide, only the even symmetric modes will be excited and a symmetric field profile is obtained by the linear combinations of the even modes [34]. For the symmetric field interference, the light from the single-mode waveguide firstly diverges in the first half and focus at the middle-central position where the replicated optical spot size from the single-mode waveguide is much smaller than the MMI waveguide cross-section. Then the light diverges in the next half of the MMI section and recouples to the single-mode waveguide. The self-imaging property in the MMI structure proves to be robust where the lateral confinement is relieved in the silica waveguide crossing design, which is quite suitable for the waveguide crossing design [35].

A typical MMI silicon waveguide crossing consists of four single-mode arms and two multimode waveguides supporting both TE_0 and TE_2 modes, as shown in Figure 3a. The tapers connecting single-mode and multimode waveguides are usually employed to reduce the back-reflection at the intersection region. The insertion loss and crosstalk of the tapered MMI silicon waveguide crossing are ~0.4 and <−30 dB, respectively. It has a footprint of 13×13 μm^2 [36]. In addition, the length of the MMI core can be reduced to less than 6 μm with a sophisticated taper by matching the Gaussian beam pattern with 0.21 dB insertion loss and −44.4 dB crosstalk [37]. Another method to fulfill the phase difference requirements between the two lowest even modes is to employ three cascaded multimode tapers, and the insertion loss is 0.13 dB and the crosstalk is −43.5 dB at the footprint of 4.16×4.16 μm^2 [38], as shown in Figure 3b. The symmetric MMI crossing supporting TE_0, TE_2 and TE_4 modes with wider MMI waveguides and the synthesized Gaussian-like focusing pattern mode has an extremely low insertion loss of 0.007 dB ± 0.004 dB and crosstalk of <−40 dB at a footprint of 30×30 μm^2 [39]. The low-loss Bloch waves can be viewed as the combination of the multimode self-focusing property with the matching of the field pattern and dielectric structure periodicities, which have low insertion loss of 0.045 dB and crosstalk of −34 dB [40]. However, the Bloch wave waveguide crossings have low fabrication tolerance and the mismatched periods can cause 0.65 dB loss, which is 15 times larger than that of the matched periods. The self-imaging property can also be employed in the silicon-based slot–waveguide crossing with insertion loss of 0.086 dB and crosstalk of −27.51 dB [41]. Similar to the shaped taper waveguide crossing, the crosstalk of MMI waveguide crossing with 110° crossing angle can be improved by more than 14 dB compared with the conventional waveguide crossing with 90° crossing angle [42]. On the other hand, the multiple ports silicon star-like crossings can greatly improve the system capacity compared with the traditional 2×2 crossing design [43–45]. The MMI-based star-like 3×3, 4×4, 5×5 and 6×6 silicon crossings are proposed and have very low insertion loss and crosstalk for all propagation channels [46]. MMI waveguide crossing is robust against the fabrication errors because the large waveguide width in MMI design is very suitable for the mature DUV technique. MMI waveguide crossing is the most popular waveguide crossing design in PICs for industry. However, the device footprint is relatively larger compared with other crossing designs, which limits its applications for ultra-compact PICs.

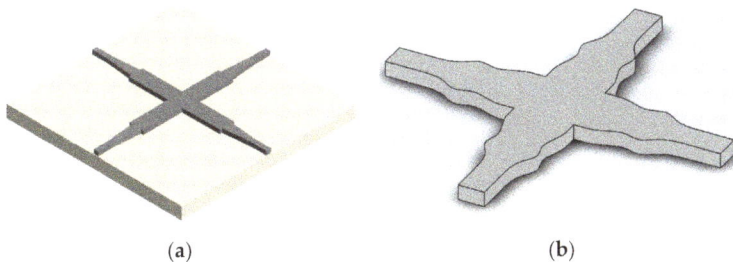

(a)   (b)

Figure 3. Schematics of (**a**) the traditional multimode interference (MMI) silicon waveguide crossing with taper transition and (**b**) the three cascaded multimode tapers silicon waveguide crossing.

2.3. Sub-Wavelength Grating Method

Sub-wavelength grating (SWG) waveguide is a periodic arrangement of two different materials having a period that is shorter than the wavelength of light. The electromagnetic wave propagation in SWGs structures with dielectric and metal layers have been theoretically studied since 1943 [47,48]. The light propagation in the SWG waveguide can be divided into three operation regimes, including reflection regime, diffraction regime and subwavelength regime. The periodic waveguides are often employed as distributed Bragg reflector (DBR) lasers in reflection regime while as grating couplers in diffraction regime [49–51]. In the subwavelength regime, the diffraction and reflection effects are eliminated and the SWG waveguide can be treated as an equivalent homogeneous medium with an approximately effective refractive index given by Rytov, which is widely used in SWG waveguide crossing design.

For SOI material platform, the SWG waveguide consists of a periodic arrangement of silicon and silica (air) layers and the effective refractive index can be tuned by chirping the duty cycle, pitch and tapering the width of the grating segments, as shown in Figure 4a. In the SWG waveguide, Bloch–Floquet optical mode is excited and its unique delocalizing property results in the efficient waveguide crossing [52]. The modal optical confinement is partly maintained for the SWG unique structure at the intersection region and the diffracting loss is reduced compared with the direct waveguide crossing [53]. On the other hand, the effective refractive index of SWG waveguide is smaller than that of the conventional silicon channel, and the scattering portion is not as strong as that in the relatively low refractive index waveguide crossing. SWG is intrinsically birefringent and the refractive indexes of the parallel and perpendicular directions are different, which can be beneficial for designing a polarization-insensitive waveguide crossing for both TE and TM modes. For the SWG waveguide crossing design, it has impressive insertion losses of 0.02 and 0.04 dB for TE and TM polarizations, respectively, and crosstalk below −40 dB with single etch fabrication step [54]. To reduce the mode mismatch introduced loss and prevent back reflection at the interface, SWG taper is often used to couple light from a conventional strip waveguide to the SWG waveguide by gradually reducing the waveguide width for fulfilling the effective index matching condition [55], as illustrated in Figure 4b. However, the structure requires ~10 × 10 μm² large adiabatic taper and an induced 0.3 dB loss per taper, which are the main drawbacks of this design. In another point of view, the SWG can be served as a refractive index engineering method and is flexible to be adopted in the mature crossing method, such as MMI crossing. E-beam lithography (EBL) is often employed due to the very small structures in SWG. EBL works by directing a beam of electros at the wafer exposing an EBL resist point by point, which is much slower than the DUV technique. The silicon waveguide crossings using the lateral index-engineered cascaded multimode-interference couplers are proved to have <0.01 dB insertion loss and <−40 dB crosstalk [56]. Owning to the compact size of MMI crossing, the index-engineered MMI coupler waveguide crossing can further reduce the footprint to around ~3 × 3 μm².

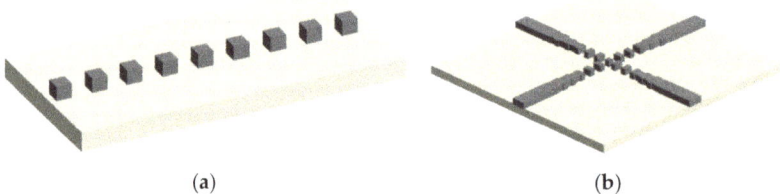

(a)   (b)

Figure 4. (a) Schematic of the sub-wavelength grating (SWG) silicon waveguide and (b) top view of SWGs silicon waveguide crossing with tapers.

2.4. Holey Subwavelength Grating Method

In recent years, numerical optimization methods are becoming more and more powerful and the single-layer inverse-designed structure is popular because the device can be easily divided into m × n pixels. Each pixel can have the assumed shapes, like square or circle, and the topology optimization process is to decide the presence or absence of these silicon pixels [57]. Figure 5a depicts the conventional inverse designed silicon waveguide crossing. To prevent the optical power diffracting into the transverse ports, the center of the intersection region and four arms are occupied by arrayed circular holes. The optical mode in the input port belongs to the resonant mode and the diffraction loss will be suppressed according to the phenomenon of resonant tunneling through a cavity with the insertion loss of ~0.2 dB [58,59]. However, the optical wavelength range is so restricted and it is impractical for the real applications. Different from the traditional etched holes, the four etched lens-like structures are placed before the intersection and form a waveguide guiding region in the crossing section, resulting in high power transmission and low crosstalk [60]. The device has a low insertion loss of <0.175 dB and low crosstalk <−37 dB with an extremely small footprint of ~1 × 1 μm², which is the most compact silicon waveguide crossing to the best of our knowledge, as shown in Figure 5b.

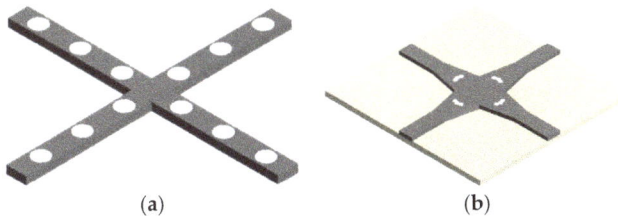

Figure 5. (a) Schematic of the inverse-designed silicon waveguide crossing and (b) top view of the ultra-compact silicon waveguide crossing with holey SWG grating method.

The inverse-designed waveguide crossing is a refractive index engineering method and can modify the refractive effective index distribution of the device as wanted. In the design [61], the silicon waveguide crossing has a square-assembling pattern with 51 × 51 pixels and the dimension of each pixel is 100 × 100 nm². The expensive EBL technique is required due to the ultra-small pixel size. The advanced algorithm determines the states of each pixel and engineer the refractive index distribution. The waveguide crossing has an insertion loss of 0.1–0.3 dB and crosstalk of <−3 dB at a footprint of 5 × 5 μm². From the large scales of optical channels to intercross connects, like 6 × 6 crossing, the utilization of the traditional 2 × 2 waveguide crossing will be very inefficient and take up huge footprint. Some multiple-input multiple-output (MIMO) crossing designs have been proposed to increase the density of the ports in a given area. The 4 × 4 PhC star-like inverse-designed waveguide crossing is proposed with a nonlinear direct-binary-search (DBS) optimization algorithm, and the insertion loss is 0.75 ± 0.2 dB and the crosstalk is <−20 dB [62].

2.5. Vertical Directional Coupler Method

Different from the previous methods of realizing the efficient silicon waveguide crossing in a single layer, the vertical directional coupler method is used to couple light from the base silicon channel to the upper or lower optical waveguide and transfer the power back into the original silicon waveguide without coming through the "dangerous" crossing region. The vertical directional coupler method, also named the waveguide bridge method, has the lowest crosstalk in theory and the key consideration for this design is to realize the efficient power coupling in the limited device footprint together with an acceptable fabrication cost.

For the silicon and silica-based waveguides, the large difference in propagation constants is difficult for transferring the optical power between these two layers [63], which results in the inefficient direction coupler and causes a large device footprint. Vertical optical waveguide coupler consisting of SOI and amorphous silicon is proposed to solve this issue, and it has the insertion loss of 0.2 dB [64] in reasonable transition length while the fabrication cost is unacceptable. For the integration of silicon and polymer layers, the optical power is coupled up and down when passing through the silicon waveguide crossings and the performances are much better with 0.08 dB insertion loss and -70 dB crosstalk [65], as shown in Figure 6a. However, the dimension of polymer waveguide is much larger than the silicon waveguide due to the weak optical confinement and the bridge waveguide crossing takes up more space compared with the crossings in a single layer. The main drawback is that the laterally stacked three silicon optical waveguides are very complicated to fabricate and these fabrication techniques are not wildly used in the mature CMOS process. Silicon nitride (SiN) is probably the most promising material for the integration with the SOI platform due to its excellent CMOS fabrication compatibility and low propagation loss in the optical communication band. The SiN over Si bridge waveguide crossing has an extremely low insertion loss of -49 dB and crosstalk of -65 dB [66], as shown in Figure 6b. For multilayer SiN-on-Si integrated photonic platforms, bilevel and trilevel grating couplers are used in these three layers platforms and they have been demonstrated to have low-loss interlayer insertion loss and ultralow-loss crosstalk.

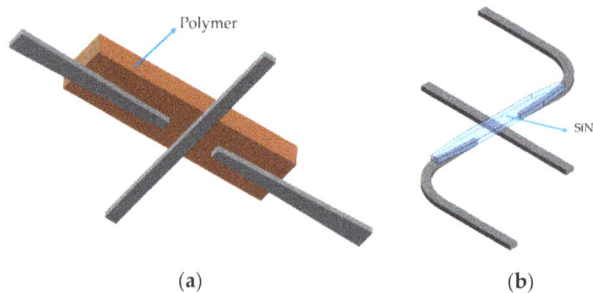

(a)   (b)

Figure 6. Schematics of the silicon waveguide crossing with vertical directional coupler consisting of (**a**) silicon and polymer layers and (**b**) silicon and SiN layers.

The low-loss, low-crosstalk, compact and low-cost integrated silicon waveguide crossings are designed to satisfy the increasing demands of PICs in the coming era of the "BIG DATA" and the "Internet of Things" [67,68]. Table 1 summarizes the typical results of different silicon waveguide crossing design methods and includes the key information about the insertion loss, crosstalk, device footprint, silicon platform thickness and fabrication cost. These sophisticated silicon waveguide crossings are proposed to fulfill all these figure of merits (FOMs) and each structure has its unique advantages and disadvantages in these FOMs. For example, the vertical directional coupler waveguide crossing has probably the best crosstalk and insertion loss performance but is not popular for its expensive fabrication cost. In addition, the MMI waveguide crossing is widely acknowledged for its fabrication tolerance and moderate insertion loss and crosstalk performances, while it is limited by its relatively large device footprint. Different silicon waveguide crossing designs are employed in different PICs.

Table 1. Comparisons of different silicon waveguide crossing designs.

Type	Institute	Insertion loss (dB)	Crosstalk (dB)	Footprint (μm²)	Thick (nm)	Fabrication Cost	Ref.
Shaped taper	YNU[1]	<0.1	<−30	7.2×1.5	320	Low	[18]
Shaped taper	Ghent Uni.	0.16	−40	6×6	220	Low	[28]
Shaped taper	UPV[2]	<0.2	<−40	6×6	250	Medium	[29]
Shaped taper	Univ. of Delaware	~0.028	<−37	9×9	220	Medium	[30]
MMI	HKUST[3]	~0.4	−30	13×13	340	Low	[36]
MMI	NCTU[4]	0.13	−43.5	4.1×4.16	220	Low	[38]
MMI	Huawei	~0.007	<−40	30×30	220	Low	[39]
MMI	Southeast Uni.	0.086	−35.58	~16×16	250	Low	[41]
SWG	NRC[5]	0.023	<−40	~10×10	260	Medium	[54]
SWG	UT Austin[6]	~0.02	<−40	~3×3	250	Medium	[56]
Holey SWG	SYSU Uni.	0.1~0.3	<−35	~5×5	220	High	[61]
Holey SWG	HUST[7]	0.75	<−20	–	220	High	[62]
Vertical DC	ISP SB RAS[8]	0.08	−70	–	–	High	[65]
Vertical DC	SNL[9]	0.16	−49	–	–	High	[66]

1. Yokohama National University. 2. Valencia Nanophotonics Technology Center. 3. Hong Kong University of Science and Technology. 4. National Chiao Tung University. 5. National Research Council, Ottawa, Canada. 6. University of Texas at Austin. 7. Huazhong University of Science and Technology. 8. Rzhanov Institute of Semiconductor Physics of the Siberian Branch of the RAS, Russian. 9. Sandia National Laboratories, Albuquerque, USA.

3. The Future Trends of Silicon Waveguide Crossing

As there is an increasing demand for ultra-high capacity optical interconnects, different silicon-based multiplexing technologies have been investigated to achieve a high-data rate [69,70]. Polarization-division-multiplexing (PDM) and mode-division-multiplexing (MDM) technologies are becoming hot research topics, because they make it possible to increase the optical link capacity by introducing dual polarization states as well as several mode states in the optical channel [71]. Polarization dependent loss (PDL) and the mode dependent loss (MDL) become additional FOMs for the silicon waveguide crossing design. Polarization and mode insensitive silicon waveguide crossings are the future trends and have great potential to be employed in the dense photonic integrated systems. Table 2 summarizes the FOMs of silicon waveguide crossing designs with multiplexing technologies and detailed characteristics will be discussed in the following paragraphs.

Table 2. Comparisons of silicon crossing with mode-division-multiplexing (MDM) and polarization-division-multiplexing (PDM) technologies.

Type	Institute	Insertion Loss (dB)	Crosstalk (dB)	Footprint (μm²)	Thick (nm)	Ref.
MDM	HUST[1]	TE_0: ~1.82 TE_1: ~0.46	<−18	21×21	220	[72]
MDM	HUST	TE_0: 0.87 TE_1: 0.54	<−50	33.7×33.7	220	[73]
MDM	Zhejiang Uni.	TM_0: 0.56 TM_1: 0.84	<−20	~32×32	340	[74]
PDM	Zhejiang Uni.	TE_0: 1.2 TM_0: 1.5	<−25	23×23	220	[75]
PDM	CUHK[2]	TE_0: 0.2 TM_0: 0.5	<−28	6×6	250	[76]
PDM	TBSI[3]	TE_0: 0.67 TM_0: 0.69	<−20	3.6×3.6	340	[77]

1. Huazhong University of Science and Technology. 2. Chinese University of Hong Kong. 3. Tsinghua-Berkeley Shenzhen Institute, Tsinghua University.

3.1. Mode-Division-Multiplexing Technology

Most traditional PICs are designed only for the fundamental optical mode for its characteristic of low loss and crosstalk. Higher order modes are attracting much more attention and are introduced to carry signals together with the fundamental modes to increase the link capacity for the PICs. The mode-division-multiplexing technology (MDM) is becoming more and more popular [78–81].

The sophisticated waveguide crossings used for the single-mode cannot be directly applied for higher order modes due to the different refractive indexes and optical power distributions, while the design principles and process may be similar to those in the single-mode waveguide crossing.

The most straightforward method is to separate the combined modes into different paths by using the mode splitter device, so the design of the single-mode waveguide crossing can be utilized in each optical path. Figure 7a shows the mode-multiplex silicon waveguide crossing with a Y-junction-based mode splitter. The TE_0 and TE_1 modes are separated and the traditional MMI silicon waveguide crossing is selected to form the efficient 2×2 crossing matrix [72]. The MMI length needs to be optimized to get a good balance between TE_0 and TE_1 modes, since there exists a gap between the beat lengths for those two modes. The device has the insertion loss of 1.82 (0.46) dB for TE_0 (TE_1) mode at 1550 nm and crosstalk of −18 dB with the device length of 23 μm. Another kind of efficient mode splitter is the PhC-assisted subwavelength asymmetric Y-junction structure and it can separate the TE_0, TE_1 and TE_2 modes simultaneously [82,83]. Similarly, a 3×3 crossing matrix is achieved by the MMI waveguide crossing whose insertion loss is less than 0.9 dB and crosstalk is lower than −24 dB for all the three modes at the length of 34 μm [84]. The other category of MDM silicon waveguide crossing is not to separate the combined modes but to manipulate the optical modes in the same crossing design. The launched TM_0 (TM_1) mode is converted to the combination of TM_0 (TM_1) and TM_2 (TM_3) mode after the tapered waveguide and self-imaging property is applied to form the efficient waveguide crossing, as shown in Figure 7b. The drawback of this MMI is that its length usually expands to the least common multiple (LCM) between these two beat lengths, so it is hard to get good transmission performances for both TM_0 and TM_1 modes. This device has an insertion of around 1.5 dB and crosstalk of −18 dB for TM_0 and TM_1 modes at a length of larger than 55 μm. The same MDM crossing design can be employed for TE_0 and TE_1 modes. The combination of strip and rib waveguides can compensate the gap of the beat length for the two modes, and reduce the device footprint with the cost of double etch fabrication. The insertion loss is <0.87 (0.54) dB for TE_0 (TE_1) mode and the crosstalk is less than −50 dB for both modes at the device length of 33.7 μm. Different from the combination of strip and rib waveguides, the subwavelength MMI crossing can realize the identical beat length for TE_0 and TE_1 mode by engineering the refractive index with the inverse design method to get the complex refractive index distribution [85]. The device has an insertion of less than 0.6 dB and crosstalk of −42 dB for both TE_0 and TE_1 modes at a length of 4.8 μm, which is a very compact device footprint for the MMI-based crossing. This design has large scalability and can be redesigned to support more modes in the future. Like the polarization-multiplexing technology, the designing process of mode-division-multiplexing technology can also be viewed as a further step compared to the traditional silicon waveguide crossing design, since more than one optical mode needs to be considered.

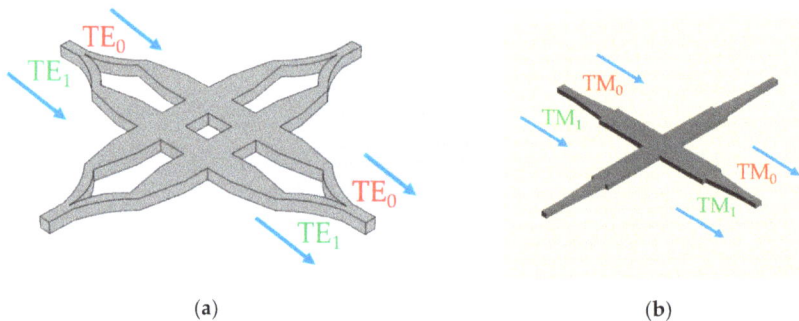

(a)   (b)

Figure 7. (**a**) Schematic of the mode-multiplexed silicon waveguide crossing with Y-junction mode splitter for fundamental transverse-electric (TE_0) and first-order transverse-electric (TE_1) mode; (**b**) top view of the mode-multiplexed silicon waveguide crossing with multimode interference structure for fundamental transverse-magnetic (TM_0) and first-order transverse-magnetic (TM_1) mode.

3.2. Polarization-Division-Multiplexing Technology

TE mode is general used in silicon channel application due to its ultra-small bending radius as small as 2.5 μm, while a few tens of micrometers are needed for TM-like mode [86]. Polarization splitters and rotators (PSRs) are used in the polarization diversity systems [87–89]. For the previously proposed waveguide crossing designs, they are more focused on TE-like mode while TM-like mode has not attracted much attention. It is intuitive that PDM technology with dual polarizations is capable of doubling the link capacity in silicon waveguide, since optical polarization states can be intrinsically divided into two perpendicular polarization states, TE and TM polarizations [90]. However, the high structural birefringence of SOI platform introduces a large effective refractive index gap and an imbalanced performance of TE and TM polarizations in the previous proposed silicon waveguide crossing designs.

For the MMI silicon waveguide crossing design, the length of the MMI section is mainly determined by the optical beat length which depends on the effective refractive index of the optical mode. Unfortunately, the beat lengths for TE and TM polarizations are not the same and a trade-off needs to be made by expanding the device into the position near the LCM to get a good transmission performance for dual polarizations, as shown in Figure 8a. The polarization-insensitive silicon MMI crossing is proposed and has the insertion loss of 0.73 dB (0.65 dB) and crosstalk of −30 dB (−45 dB) for TE (TM) polarization with the device length of 23×23 μm^2 at 220 nm SOI platform, which is a relatively large device footprint in the modern PICs. The polarization-insensitive silicon shaped taper waveguide crossing is demonstrated by using the numerical inverse design method and the device can achieve insertion loss of 0.08 (0.07) dB and the crosstalk of −32 (−35) dB for TE (TM) mode with the device length of 6 μm at 250 nm SOI platform. Previously, we have explained the theoretical principle of tuning the effective refractive index with the subwavelength grating method. However, the taper will introduce the additional insertion loss. We proposed a SWG-assisted MMI crossing and reduced the MMI-based device footprint by narrowing the refractive index gap between TE and TM polarizations with the SWGs structure [91]. The device has the insertion loss of 0.69 (0.61) dB and crosstalk of −45 (−35) dB for TE (TM) polarization at the moderate footprint of 12.5×12.5 μm^2 on 220 nm SOI platform. For the holey SWG silicon waveguide crossing, the PDL is also taken into consideration as another designing targets, as shown in Figure 8b. The device can achieve the insertion loss of −0.67 (−0.69 dB) and less than −20 dB for TE (TM) mode at the extremely small footprint of 3.6×3.6 μm^2 at 340 nm SOI platform. The designing principle and process keep roughly the same whether the waveguide crossing is polarization sensitive or not. The polarization-insensitive waveguide crossing adds another design constraint since the TM polarization mode is also needed to be considered.

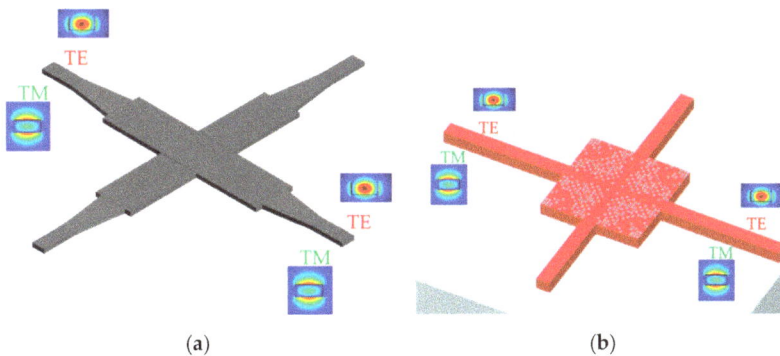

(a)   (b)

Figure 8. (**a**) Schematic of the polarization-multiplexed MMI silicon waveguide crossing; (**b**) top view of the polarization-multiplexed inverse-designed silicon waveguide crossing.

3.3. Summary

Silicon photonic waveguide crossing is a basic passive building block in PICs and the target is manipulating light to effectively go through the "dangerous" intersection region of structures, with features near or below the scale of the electromagnetic wavelength. The workflow is to design new crossing structures with some prior known physical effects in waveguide crossing, such as the optical coupling theory in taper, the self-imaging principle in multimode waveguide and so on. To meet the increasing requirements for the optical interconnect, the effective waveguide crossing is not only limited to low insertion loss and crosstalk but also needs to be integrated with different merits, such as the polarization insensitivity, mode multiplexing technology and ultra-compact footprint. On the other hand, inverse-design methods in nanophotonics are widely adopted by researchers and impressive progresses are achieved in different photonic devices [92,93].

4. Conclusions

In this paper, we reviewed the state-of-the-art and provided our perspectives on the silicon waveguide crossing. It gives the theoretical principle and device performance about several important waveguide crossing design methods such as shaped taper, multimode interference, subwavelength grating, holey subwavelength grating and vertical directional coupler. On the other hand, we also include some future trends of silicon waveguide crossing to meet the increasing requirements in data rates for PICs, like polarization-division-multiplexing and mode-division-multiplexing technologies. We believe that a boost in the number of silicon photonic products is coming to the market and an increase in the number of complex silicon photonic systems is being developed in both academia and industry. The silicon waveguide crossing is one of the basic building blocks in PICs and its development will pave the path of complex and functional PICs in the near future.

Author Contributions: S.W., X.M., L.C., S.M. and H.Y.F. contributed to the writing, reviewing and editing of this paper. All authors have read and agreed to the published version of the manuscript.

Funding: This research work is supported by Shenzhen Science and Technology Innovation Commission (Project: JCYJ20180507183815699, JCYJ20170818094001391, KQJSCX20170727163424873), Tsinghua-Berkeley Shenzhen Institute (TBSI) Faculty Start-up Fund and Shenzhen Data Science and Information Technology Engineering Laboratory.

Conflicts of Interest: The authors declare no conflicts of interest.

References

1. Fang, Z.; Zhao, C.Z. Recent Progress in Silicon Photonics: A Review. *ISRN Opt.* **2012**, *2012*, 1–27. [CrossRef]
2. Waldrop, M.M. More than moore. *Nature* **2016**, *530*, 144–148. [CrossRef]
3. Photonic-Electronic Integrated Circuits. Available online: https://www.edmundoptics.com/resources/trending-in-optics/photonic-electronic-integrated-circuits/ (accessed on 8 February 2020).
4. Doylend, J.K.; Knights, A.P. The evolution of silicon photonics as an enabling technology for optical interconnection. *Laser Photonics Rev.* **2012**, *6*, 504–525. [CrossRef]
5. Smit, M.; van der Tol, J.; Hill, M. Moore's law in photonics. *Laser Photonics Rev.* **2012**, *6*, 1–13. [CrossRef]
6. Zou, Y.; Chakravarty, S.; Chung, C.J.; Xu, X.; Chen, R.T. Mid-infrared silicon photonic waveguides and devices. *Photonics Res.* **2018**, *6*, 254–276. [CrossRef]
7. Hu, T.; Dong, B.; Luo, X.; Liow, T.Y.; Song, J.; Lee, C.; Lo, G.Q. Silicon photonic platforms for mid-infrared applications. *Photonics Res.* **2017**, *1*, 417–430. [CrossRef]
8. Lin, H.; Luo, Z.; Gu, T.; Kimerling, L.C.; Wada, K.; Agarwal, A.; Hu, J. Mid-infrared integrated photonics on silicon: A perspective. *Nanophotonics* **2017**, *7*, 393–420. [CrossRef]
9. Ram, R.J. CMOS Photonic Integrated Circuits. In Proceedings of the 2012 Optical Fiber Communication Conference, Los Angeles, CA, USA, 4–8 March 2012.
10. Soref, R. The Past, Present, and Future of Silicon Photonics. *IEEE J. Sel. Top. Quantum Electron.* **2006**, *12*, 1678–1687. [CrossRef]

11. Stojanović, V.; Ram, R.J.; Popović, M.; Lin, S.; Moazeni, S.; Wade, M.; Sun, C.; Alloatti, L.; Atabaki, A.; Pavanello, F.; et al. Monolithic silicon-photonic platforms in state-of-the-art CMOS SOI processes [Invited]. *Opt. Express* **2018**, *26*, 13106–13121.

12. Rickman, A. The commercialization of silicon photonics. *Nat. Photonics* **2014**, *8*, 579. [CrossRef]

13. Soref, R.; Lorenzo, J. Single-crystal silicon: A new material for 1.3 and 1.6 μm integrated-optical components. *Electron. Lett.* **1985**, *21*, 953–954. [CrossRef]

14. Thomson, D.; Zilkie, A.; Bowers, J.E.; Komljenovic, T.; Reed, G.T.; Vivien, L.; Marris-Morini, D.; Cassan, E.; Virot, L.; Fédéli, J.M.; et al. Roadmap on silicon photonics. *J. Opt.* **2016**, *18*, 73003. [CrossRef]

15. Wen, K.; Rumley, S.; Samadi, P.; Chen, C.P.; Bergman, K. Silicon photonics in post Moore's Law era: Technological and architectural implications. In Proceedings of the 2016 Conference on International Workshop on Post Moore's Era Supercomputing, Salt Laker City, UT, USA, 11–14 November 2016; p. 12.

16. Watari, T.; Murano, H. Packaging Technology for the NEC SX Supercomputer. *IEEE Trans. Compon. Hybrids Manuf. Technol.* **1985**, *8*, 462–467. [CrossRef]

17. Bogaerts, W.; Fiers, M.; Dumon, P. Design challenges in silicon photonics. *IEEE J. Sel. Top. Quantum Electron.* **2013**, *20*, 1–8. [CrossRef]

18. Fukazawa, T.; Hirano, T.; Ohno, F.; Baba, T. Low Loss Intersection of Si Photonic Wire Waveguides. *Jpn. J. Appl. Phys.* **2004**, *43*, 646–647. [CrossRef]

19. Sherwood-Droz, N.; Wang, H.; Chen, L.; Lee, B.G.; Biberman, A.; Bergman, K.; Lipson, M. Optical 4x4 hitless slicon router for optical networks-on-chip (NoC). *Opt. Express* **2008**, *16*, 15915–15922. [CrossRef]

20. Hu, T.; Qiu, H.; Yu, P.; Qiu, C.; Wang, W.; Jiang, X.; Yang, M.; Yang, J. Wavelength-selective 4×4 nonblocking silicon optical router for networks-on-chip. *Opt. Lett.* **2011**, *36*, 4710–4712. [CrossRef]

21. Sakai, A.; Hara, G.; Baba, T. Propagation characteristics of ultrahigh-Δ optical waveguide on silicon-on-insulator substrate. *Jpn. J. Appl. Phys.* **2001**, *40*, L383. [CrossRef]

22. Qiu, C.; Sheng, Z.; Li, H.; Liu, W.; Li, L.; Pang, A.; Wu, A.; Wang, X.; Zou, S.; Gan, F. Fabrication, characterization and loss analysis of silicon nanowaveguides. *J. Lightwave Technol.* **2014**, *32*, 2303–2307. [CrossRef]

23. Vlasov, Y.; McNab, S. Losses in single-mode silicon-on-insulator strip waveguides and bends. *Opt. Express* **2004**, *12*, 1622–1631. [CrossRef]

24. Liu, H.; Tam, H.; Wai, P.K.A.; Pun, E. Low-loss waveguide crossing using a multimode interference structure. *Opt. Commun.* **2004**, *241*, 99–104. [CrossRef]

25. Burns, W.K.; Milton, A.F.; Lee, A.B. Optical waveguide parabolic coupling horns. *Appl. Phys. Lett.* **1977**, *30*, 28–30. [CrossRef]

26. Fu, Y.; Ye, T.; Tang, W.; Chu, T. Efficient adiabatic silicon-on-insulator waveguide taper. *Photonics Res.* **2014**, *2*, A41–A44. [CrossRef]

27. Nelson, A.R. Coupling optical waveguides by tapers. *Appl. Opt.* **1975**, *14*, 3012–3015. [CrossRef]

28. Bogaerts, W.; Dumon, P.; Van Thourhout, D.; Baets, R. Low-loss, low-cross-talk crossings for silicon-on-insulator nanophotonic waveguides. *Opt. Lett.* **2007**, *32*, 2801–2803. [CrossRef]

29. Sanchis, P.; Villalba, P.; Cuesta, F.; Håkansson, A.; Griol, A.; Galán, J.V.; Brimont, A.; Martí, J. Highly efficient crossing structure for silicon-on-insulator waveguides. *Opt. Lett.* **2009**, *34*, 2760–2762. [CrossRef]

30. Ma, Y.; Zhang, Y.; Yang, S.; Novack, A.; Ding, R.; Lim, A.E.J.; Lo, G.Q.; Baehr-Jones, T.; Hochberg, M. Ultralow loss single layer submicron silicon waveguide crossing for SOI optical interconnect. *Opt. Express* **2013**, *21*, 29374–29382. [CrossRef]

31. Xie, Y.; Xu, J.; Zhang, J. Elimination of cross-talk in silicon-on-insulator waveguide crossings with optimized angle. *Opt. Eng.* **2011**, *50*, 064601.

32. Tanaka, D.; Shoji, Y.; Kintaka, K.; Kawashima, H.; Ikuma, Y.; Tsuda, H. Low-crosstalk offset crossing waveguide fabricated on SOI substrates. In Proceedings of the 2010 OECC, Sapporo, Japan, 5–9 July 2010; pp. 870–871.

33. Amin, S.; Aziz, K. Multimode interference (MMI) devices: A survey. In Proceedings of the 2010 the 8th International Conference on Frontiers of Information Technology, Islamabad, Pakistan, 21–23, December 2010; pp. 1–6.

34. Soldano, L.B.; Pennings, E.C.M. Optical multi-mode interference devices based on self-imaging: Principles and applications. *J. Lightwave Technol.* **1995**, *13*, 615–627. [CrossRef]

35. Stuart, H.R. Waveguide lenses with multimode interference for low-loss slab propagation. *Opt. Lett.* **2003**, *28*, 2141–2143. [CrossRef]

36. Chen, H.; Poon, A.W. Low-Loss Multimode-Interference-Based Crossings for Silicon Wire Waveguides. *IEEE Photonics Technol. Lett.* **2006**, *18*, 2260–2262. [CrossRef]

37. Chen, C.H.; Chiu, C.-H. Taper-integrated multimode-interference based waveguide crossing design. *IEEE J. Quantum Electron.* **2010**, *46*, 1656–1661. [CrossRef]

38. Chen, C.H. Compact waveguide crossings with a cascaded multimode tapered structure. *Appl. Opt.* **2015**, *54*, 828–833. [CrossRef]

39. Dumais, P.; Goodwill, D.J.; Celo, D.; Jiang, J.; Bernier, E. Three-mode synthesis of slab Gaussian beam in ultra-low-loss in-plane nanophotonic silicon waveguide crossing. In Proceedings of the 14th International Conference on Group IV Photonics, Berlin, Germany, 23–25 August 2017; pp. 97–98.

40. Popović, M.A.; Ippen, E.P.; Kärtner, F.X. Low-loss bloch waves in open structures and highly compact, efficient Si waveguide-crossing arrays. In Proceedings of the 2007 Conference on Lasers and Electro-Optics Society Annual Meeting, Tallahassee, FL, USA, 21–25 October 2007; pp. 56–57.

41. Xu, Y.; Wang, J.; Xiao, J.; Sun, X. Design of a compact silicon-based slot–waveguide crossing. *Appl. Opt.* **2013**, *52*, 3737–3744. [CrossRef]

42. Kim, S.H.; Cong, G.; Kawashima, H.; Hasama, T.; Ishikawa, H. Low-crosstalk waveguide crossing based on 1 × 1 MMI structure of silicon-wire waveguide. In Proceedings of the 2013 Pacific Rim Conference on Lasers and Electro-Optics, Kyoto, Japan, 30 June–4 July 2013; pp. 1–2.

43. García-Meca, C.; Lechago, S.; Brimont, A.; Griol, A.; Mas, S.; Sánchez, L.; Bellieres, L.; Losilla, N.S.; Martí, J. On-chip wireless silicon photonics: From reconfigurable interconnects to lab-on-chip devices. *Light Sci. Appl.* **2017**, *6*, e17053. [CrossRef]

44. Xu, H.; Shi, Y. Metamaterial-Based Maxwell's Fisheye Lens for Multimode Waveguide Crossing. *Laser Photonics Rev.* **2018**, *12*, 10. [CrossRef]

45. Li, S.; Zhou, Y.; Dong, J.; Zhang, X.; Cassan, E.; Hou, J.; Yang, C.; Chen, S.; Gao, D.; Chen, H. Universal multimode waveguide crossing based on transformation optics. *Optica* **2018**, *5*, 1549–1556. [CrossRef]

46. Chen, D.; Wang, L.; Zhang, Y.; Hu, X.; Xiao, X.; Yu, S. Ultralow crosstalk and loss CMOS compatible silicon waveguide star-crossings with arbitrary included angles. *ACS Photonics* **2018**, *5*, 4098–4103. [CrossRef]

47. Tamm, I.E.; Ginzburg, V.L. Theory of electromagnetic processes in a layered core. *Izv. Akad. Nauk SSSR Ser. Fiz.* **1943**, *7*, 30–51.

48. Rytov, S. Electromagnetic properties of a finely stratified medium. *Sov. Phys. JEPT* **1956**, *2*, 466–475.

49. Taillaert, D.; Bienstman, P.; Baets, R. Compact efficient broadband grating coupler for silicon-on-insulator waveguides. *Opt. Lett.* **2004**, *29*, 2749–2751. [CrossRef]

50. Shyh, W. Principles of distributed feedback and distributed Bragg-reflector lasers. *IEEE J. Quantum Electron.* **1974**, *10*, 413–427. [CrossRef]

51. Halir, R.; Bock, P.J.; Cheben, P.; Ortega-Moñux, A.; Alonso-Ramos, C.; Schmid, J.H.; Lapointe, J.; Xu, D.X.; Wangüemert-Pérez, J.G.; Molina-Fernández, Í.; et al. Waveguide sub-wavelength structures: A review of principles and applications. *Laser Photonics Rev.* **2015**, *9*, 25–49. [CrossRef]

52. Bock, P.J.; Cheben, P.; Schmid, J.H.; Lapointe, J.; Delâge, A.; Janz, S.; Aers, G.C.; Xu, D.X.; Densmore, A.; Hall, T.J. Subwavelength grating periodic structures in silicon-on-insulator: A new type of microphotonic waveguide. *Opt. Express* **2010**, *18*, 20251–20262. [CrossRef] [PubMed]

53. Chen, L.R.; Wang, J.; Naghdi, B.; Glesk, I. Subwavelength Grating Waveguide Devices for Telecommunications Applications. *IEEE J. Sel. Top. Quantum Electron.* **2018**, *25*, 1–11. [CrossRef]

54. Bock, P.J.; Cheben, P.; Schmid, J.H.; Lapointe, J.; Delâge, A.; Xu, D.X.; Janz, S.; Densmore, A.; Hall, T.J. Subwavelength grating crossings for silicon wire waveguides. *Opt. Express* **2010**, *18*, 16146–16155. [CrossRef]

55. Schmid, J.H.; Cheben, P.; Janz, S.; Lapointe, J.; Post, E.; Delâge, A.; Densmore, A.; Lamontagne, B.; Waldron, P.; Xu, D.X. Subwavelength Grating Structures in Silicon-on-Insulator Waveguides. *Adv. Opt. Technol.* **2008**, *2008*, 685489. [CrossRef]

56. Zhang, Y.; Hosseini, A.; Xu, X.; Kwong, D.; Chen, R.T. Ultralow-loss silicon waveguide crossing using Bloch modes in index-engineered cascaded multimode-interference couplers. *Opt. Lett.* **2013**, *38*, 3608–3611. [CrossRef]

57. Shen, B.; Wang, P.; Polson, R.; Menon, R. An integrated-nanophotonics polarization beamsplitter with 2.4 × 2.4 μm² footprint. *Nat. Photonics* **2015**, *9*, 378–382. [CrossRef]

58. Johnson, S.G.; Manolatou, C.; Fan, S.; Villeneuve, P.R.; Joannopoulos, J.D.; Haus, H.A. Elimination of cross talk in waveguide intersections. *Opt. Lett.* **1998**, *23*, 1855–1857. [CrossRef]

59. Manolatou, C.; Steven, G.; Johnson, S.F.; Pierre, R.; Villeneuve, H.A.; Haus, J.D. High-Density Integrated Optics. *J. Lightwave Technol.* **1999**, *17*, 1682–1692. [CrossRef]

60. Han, H.L.; Li, H.; Zhang, X.P.; Liu, A.; Lin, T.Y.; Chen, Z.; Lv, H.B.; Lu, M.H.; Liu, X.P.; Chen, Y.F. High performance ultra-compact SOI waveguide crossing. *Opt. Express* **2018**, *26*, 25602–25610. [CrossRef] [PubMed]

61. Xu, P.; Zhang, Y.; Shao, Z.; Yang, C.; Liu, L.; Chen, Y.; Yu, S. 5×5 μm^2 compact waveguide crossing optimized by genetic algorithm. In Proceedings of the 2017 Conference on Asia Communications and Photonics Conference, Guangzhou, China, 10–13 November 2017; pp. 1–3.

62. Lu, L.; Zhang, M.; Zhou, F.; Chang, W.; Tang, J.; Li, D.; Ren, X.; Pan, Z.; Cheng, M.; Liu, D. Inverse-designed ultra-compact star-crossings based on PhC-like subwavelength structures for optical intercross connect. *Opt. Express* **2017**, *25*, 18355–18364. [CrossRef] [PubMed]

63. Wakayama, Y.; Kita, T.; Yamada, H. Optical crossing and integration using hybrid si-wire/silica waveguides. *Jpn. J. Appl. Phys.* **2011**, *50*, 4–20. [CrossRef]

64. Sun, R.; Beals, M.; Pomerene, A.; Cheng, J.; Hong, C.Y.; Kimerling, L.; Michel, J. Impedance matching vertical optical waveguide couplers for dense high index contrast circuits. *Opt. Express* **2008**, *16*, 11682–11690. [CrossRef]

65. Tsarev, A.V. Efficient silicon wire waveguide crossing with negligible loss and crosstalk. *Opt. Express* **2011**, *19*, 13732–13737. [CrossRef]

66. Jones, A.M.; DeRose, C.T.; Lentine, A.L.; Trotter, D.C.; Starbuck, A.L.; Norwood, R.A. Ultra-low crosstalk, CMOS compatible waveguide crossings for densely integrated photonic interconnection networks. *Opt. Express* **2018**, *21*, 12002–12013. [CrossRef]

67. Li, Z.; Shubin, I.; Zhou, X. Optical interconnects: Recent advances and future challenges. *Opt. Express* **2015**, *23*, 3717–3720. [CrossRef]

68. Vahdat, A.; Liu, H.; Zhao, X.; Johnson, C. The emerging optical data center. In Proceedings of the 2011 Conference on Optical Fiber Communication Conference, Los Angeles, CA, USA, 6–10 March 2011.

69. Li, C.; Wu, H.; Tan, Y.; Wang, S.; Dai, D. Silicon-based on-chip hybrid (de)multiplexers. *Sci. China Inf. Sci.* **2018**, *61*, 080407. [CrossRef]

70. Winzer, P.J. Making spatial multiplexing a reality. *Nat. Photonics* **2014**, *8*, 345–348. [CrossRef]

71. Dai, D.; Bowers, J.E. Silicon-based on-chip multiplexing technologies and devices for Peta-bit optical interconnects. *Nanophotonics* **2014**, *3*, 283–311. [CrossRef]

72. Sun, C.; Yu, Y.; Zhang, X. Ultra-compact waveguide crossing for a mode-division multiplexing optical network. *Opt. Lett.* **2017**, *42*, 4913–4916. [CrossRef] [PubMed]

73. Wu, B.; Yu, Y.; Zhang, X. Ultralow Loss Waveguide Crossing with Low Imbalance for Two Transverse Electric Modes. In Proceedings of the 2018 Conference on Asia Communications and Photonics, Hangzhou, China, 26–29 October 2018; pp. 1–3.

74. Xu, H.; Shi, Y. Dual-mode waveguide crossing utilizing taper-assisted multimode-interference couplers. *Opt. Lett.* **2016**, *41*, 5381–5384. [CrossRef] [PubMed]

75. Chen, J.; Shi, Y. Polarization-insensitive silicon waveguide crossing based on multimode interference couplers. *Opt. Lett.* **2018**, *43*, 5961–5964. [CrossRef] [PubMed]

76. Yu, Z.; Feng, A.; Xi, X.; Sun, X. Inverse-designed low-loss and wideband polarization-insensitive silicon waveguide crossing. *Opt. Lett.* **2019**, *44*, 77–80. [CrossRef]

77. Wu, S.; Mu, X.; Cheng, L.; Tu, X.; Fu, H.Y. Inverse-designed Compact and Polarization-insensitive Waveguide Crossing. In Proceedings of the 2019 Conference on Asia Communications and Photonics, Chengdu, China, 2–5 November 2019; p. M4A-280.

78. Li, C.; Liu, D.; Dai, D. Multimode silicon photonics. *Nanophotonics* **2019**, *8*, 227–247. [CrossRef]

79. Wang, J.; Chen, S.; Dai, D. Silicon hybrid demultiplexer with 64 channels for wavelength/mode-division multiplexed on-chip optical interconnects. *Opt. Lett.* **2014**, *39*, 6993–6996. [CrossRef]

80. Koonen, A.M.J.; Chen, H.; van den Boom, H.P.; Raz, O. Silicon Photonic Integrated Mode Multiplexer and Demultiplexer. *IEEE Photonics Technol. Lett.* **2012**, *24*, 1961–1964. [CrossRef]

81. Stern, B.; Zhu, X.; Chen, C.P.; Tzuang, L.D.; Cardenas, J.; Bergman, K.; Lipson, M. On-chip mode-division multiplexing switch. *Optica* **2015**, *2*, 530–535. [CrossRef]

82. Chang, W.; Lu, L.; Ren, X.; Li, D.; Pan, Z.; Cheng, M.; Liu, D.; Zhang, M. Ultra-compact mode (de) multiplexer based on subwavelength asymmetric Y-junction. *Opt. Express* **2018**, *26*, 8162–8170. [CrossRef]

83. Liu, Y.; Xu, K.; Wang, S.; Shen, W.; Xie, H.; Wang, Y.; Xiao, S.; Yao, Y.; Du, J.; He, Z.; et al. Arbitrarily routed mode-division multiplexed photonic circuits for dense integration. *Nat. Commun.* **2019**, *10*, 1–7. [CrossRef] [PubMed]

84. Chang, W.; Lu, L.; Ren, X.; Lu, L.; Cheng, M.; Liu, D.; Zhang, M. An Ultracompact Multimode Waveguide Crossing Based on Subwavelength Asymmetric Y-Junction. *IEEE Photonics J.* **2018**, *10*, 1–8. [CrossRef]

85. Chang, W.; Lu, L.; Ren, X.; Li, D.; Pan, Z.; Cheng, M.; Liu, D.; Zhang, M. Ultracompact dual-mode waveguide crossing based on subwavelength multimode-interference couplers. *Photonics Res.* **2018**, *6*, 660–665. [CrossRef]

86. Yamada, K. Silicon photonic wire waveguides: Fundamentals and applications. In *Silicon Photonics II*; Springer: Berlin/Heidelberg, Germany, 2011; pp. 1–29.

87. Ding, Y.; Liu, L.; Peucheret, C.; Ou, H. Fabrication tolerant polarization splitter and rotator based on a tapered directional coupler. *Opt. Express* **2012**, *20*, 20021–20027. [CrossRef] [PubMed]

88. Vermeulen, D.; Selvaraja, S.; Verheyen, P.; Absil, P.; Bogaerts, W.; Van Thourhout, D.; Roelkens, G. Silicon-on-Insulator Polarization Rotator Based on a Symmetry Breaking Silicon Overlay. *IEEE Photonics Technol. Lett.* **2012**, *24*, 482–484. [CrossRef]

89. Liu, L.; Ding, Y.; Yvind, K.; Hvam, J.M. Silicon-on-insulator polarization splitting and rotating device for polarization diversity circuits. *Opt. Express* **2011**, *19*, 12646–12651. [CrossRef]

90. Dai, D.; Bauters, J.; Bowers, J.E. Passive technologies for future large-scale photonic integrated circuits on silicon: Polarization handling, light non-reciprocity and loss reduction. *Light Sci. Appl.* **2012**, *1*, e1. [CrossRef]

91. Wu, S.; Mu, X.; Cheng, L.; Tu, X.; Fu, H.Y. Polarization-insensitive Waveguide Crossings Based on SWGs-assisted MMI. In Proceedings of the 2019 Conference on 18th Optical Communications and Networks, Huangshan, China, 5–8 August 2019; pp. 1–3.

92. Molesky, S.; Lin, Z.; Piggott, A.Y.; Jin, W.; Vucković, J.; Rodriguez, A.W. Inverse design in nanophotonics. *Nat. Photonics* **2018**, *12*, 659–670. [CrossRef]

93. Sajedian, I.; Badloe, T.; Rho, J. Finding the best design parameters for optical nanostructures using reinforcement learning. *arXiv* **2018**, arXiv:1810.10964. Available online: https://arxiv.org/abs/1810.10964 (accessed on 20 March 2020).

micromachines

MDPI

Article

Effect of the Thermal History on the Crystallinity of Poly (L-lactic Acid) During the Micromolding Process

Hiroaki Takehara [1,2,*], Yuki Hadano [1], Yukihiro Kanda [1] and Takanori Ichiki [1,2]

[1] Department of Materials Engineering, School of Engineering, The University of Tokyo, 7-3-1 Hongo, Bunkyo-ku, Tokyo 113-8656, Japan; hadano@bionano.t.u-tokyo.ac.jp (Y.H.); kanda@bionano.t.u-tokyo.ac.jp (Y.K.); ichiki@bionano.t.u-tokyo.ac.jp (T.I.)

[2] Innovation Center of NanoMedicine (iCONM), 3-25-14 Tonomachi, Kawasaki, Kanagawa 210-0821, Japan

* Correspondence: takehara@bionano.t.u-tokyo.ac.jp; Tel.: +81-3-5841-7781

Received: 29 February 2020; Accepted: 22 April 2020; Published: 25 April 2020

Abstract: The micromolding process using biocompatible thermoplastic polymers is highly attractive as a fabrication process of microdevices in biomedical applications. In this study, we investigated the effect of the thermal history in the micromolding process on the crystallinity of semi-crystalline polymers, such as poly (L-lactic acid) (PLLA), during their crystallization from the amorphous and molten states. In particular, the thermal history in the micromolding process using poly(dimethylsiloxane) replica mold embedded with a thermocouple was recorded. The crystallinity of PLLA constructs fabricated using the micromolding process was measured via wide-angle X-ray scattering, and crystallization kinetics was analyzed based on the Kolmogorov–Johnson–Mehl–Avrami equation. A crystallization rate of $k = 0.061$ min^{-n} was obtained in the micromolding process of PLLA crystallization from the amorphous state, accompanied by the quenching operation, forming a large number of crystal nuclei. Finally, the fabrication of PLLA microneedles was performed using micromolding processes with different thermal histories. The information about the thermal history during the micromolding process is significant in the development of polymer microdevices to achieve better material properties.

Keywords: semi-crystalline polymer; PLLA; thermoplastics; microdevice

---

1. Introduction

Bioabsorbable polymers are useful as materials for constructing medical devices used in the human body. Furthermore, the development of precision processing and microfabrication techniques for device miniaturization is in progress. Poly (lactic acid) polymers such as PLLA are approved by the US Food and Drug Administration (FDA) as generally recognized as safe (GRAS) [1]. Therefore, poly (L-lactic acid) (PLLA) is one of the most promising thermoplastic materials for biomedical applications, including surgical sutures, implants, and microneedles [2,3].

The micromolding process is an established manufacturing technology for fabricating biomedical microdevices using thermoplastic polymers. The production cost of the micromolding process mainly relies on the fabrication cost for the master mold, and thus, the cost of the employed material is negligibly low. This aspect of the micromolding process enables the use of high-end materials (e.g., the pharmaceutical grade) even for disposable usage [4]. However, polymer materials change their material properties depending on the process conditions, especially the thermal history during the molding process [5,6].

The information regarding the fabrication process and material properties can be used as a guideline for the development of polymer microdevices to ensure their excellent material properties. In particular, semi-crystalline polymers including PLLA change their crystallinity during the micromolding process.

It has been found that the degree of crystallinity of PLLA impacts the hydrolytic degradation kinetics inside the human body in clinical applications [7]. The mechanism of the hydrolytic degradation could be considered as the combination of chemical hydrolysis of polymer chain scission at the ester bond [8] and the diffusion of water molecules and divided oligomers via the bulk-erosion mechanism [9]. The hydrolysis of PLLA predominantly occurs in the amorphous region [10] because of the difficulty of the diffusion of water molecules into the rigid crystalline region [11]. The degree of crystallinity also determines their physical properties, including mechanical behavior. The mechanism of the deformation of the amorphous and crystalline PLLA polymers indicates that the amorphous region could be deformed by crazing, and the crystalline region could show cavitation and fibrillated shear [12]. Crystalline PLLA materials often show a higher modulus of elasticity than amorphous PLLA materials [13]. In contrast, amorphous PLLA materials show higher bending strength than crystalline because of the plastic deformation after reaching their yield point [14].

Although previous research well-argued and disclosed the correlation between the crystallinities of PLLA and the material properties of hydrolysis and mechanical behavior, there is limited research about the implementation of the above knowledge into the micromolding process for fabricating biomedical microdevices. Harris and Lee reported that adding talc and ethylenebis-stearamide (EBS) increases the crystallization rate in the injection molding process [15]. This finding could indicate that the talc and EBS function as a physical nucleation agent, and thus, shortens the nucleation and crystal growth rate. Although such an approach is feasible to use as environmentally friendly materials, it might not be feasible in clinical use because of the necessity to guaranty the safety of the added materials. Iozzino et al. reported that crystalline regions show better resistance to hydrolysis than amorphous regions in the micromolded pure PLA biphasic samples [16]. However, the crystallization rate of the PLLA in the micromolding process and the difference in the thermal histories (i.e., crystallization processes from the amorphous state and the molten state) have not been investigated, despite the importance of this information for fabricating biomedical microdevices, such as microneedles. To obtain polymeric devices with sufficient crystallinity to control optimal physical and chemical properties, the information about the crystallization rate in the micromolding process has importance.

To understand how to control the crystallinity of polymer materials fabricated using the micromolding process, we investigated the effect of the thermal history on the crystallinity of PLLA during the micromolding process. The findings of this study can be applied to the fabrication process of microdevices for biomedical applications. The crystallization of semi-crystalline polymers during the micromolding process can be classified into two main types: (i) the crystallization from the molten state and (ii) the crystallization from the amorphous (glassy) state. This study investigates the effect of the thermal history during the two types of the crystallization process on the crystallinity and crystallization kinetics of PLLA.

2. Materials and Methods

2.1. Materials

PLLA pellet (Lot # STBH0071, Resomer L 206 S, Sigma-Aldrich Corp., St. Louis, MO, USA) and a prepolymer of poly(dimethylsiloxane) (PDMS) (Sylgard 184, Dow Corning Co., Midland, MI, USA) were used in this study. The PLLA pellet was characterized by gel-permeation chromatography (GPC) analysis. Monodisperse polystyrene standards (Sigma-Aldrich Corp., St. Louis, MO, USA) were used for calibration using GPC KF-804L column (Shodex, Showa Denko K.K., Tokyo, Japan) with tetrahydrofuran (THF, Fujifilm Wako Pure Chemical Corp., Tokyo, Japan) as an eluent at 40 °C and a flow rate of 1 mL/min. The molecular weight (M_w) and polydispersity index (M_w/M_n) of the PLLA pellet were evaluated as $M_w = 12.1$ kg/mol and $M_w/M_n = 1.34$, respectively. The PDMS prepolymer was prepared using the mixing ratio of BASE:CAT = 10:1.5. To form molds for the micromolding process, the prepolymer solution was cured at 85 °C for 2 h.

2.2. Sample Preparation with Different Thermal History in the Micromolding Process

The change in the crystallinity of the PLLA due to the temperature change during molding was examined using a PLLA flat plate sample. Figure 1 shows the PLLA micromolding process, including its thermal history. The PLLA pellet was melted on a PDMS mold and molded into the PLLA flat plate sample (width 20 mm, height 20 mm, thickness 0.1 mm). The process temperature was measured using a thermocouple (TCTG022, Sakaguchi E.H VOC Corp., Tokyo, Japan) embedded inside the PDMS mold. To record the thermal history during the micromolding process, the thermocouple was connected to a thermometer (BAT-10, Physitemp Instruments LLC, Clifton, NJ, USA) and a multimeter (7461A, ADCMT Corp., Tokyo, Japan).

Crystallization of semi-crystalline polymers can be classified into two processes of crystallization from the molten and amorphous states. Therefore, micromolding processes with two different thermal histories were performed (referred to as process (i) and process (ii)). In process (i), the mold was maintained at the crystallization temperature (120–130 °C) for 2.5, 5.0, 7.5, 10.0, 12.5, 15, and 20 min after PLLA melting and then cooled to room temperature. In process (ii), the mold was rapidly cooled after PLLA melting to 40–50 °C, which is below T_g of PLLA, and then heated to the crystallization temperature (120–130 °C) and held for the crystallization time $t = 3.0, 4.0, 5.0, 7.5, 10,$ and 15 min. For preparing the samples with a crystallization time of 0 min, the mold was rapidly cooled (>200 °C/min) to room temperature (20–25 °C) after PLLA melting.

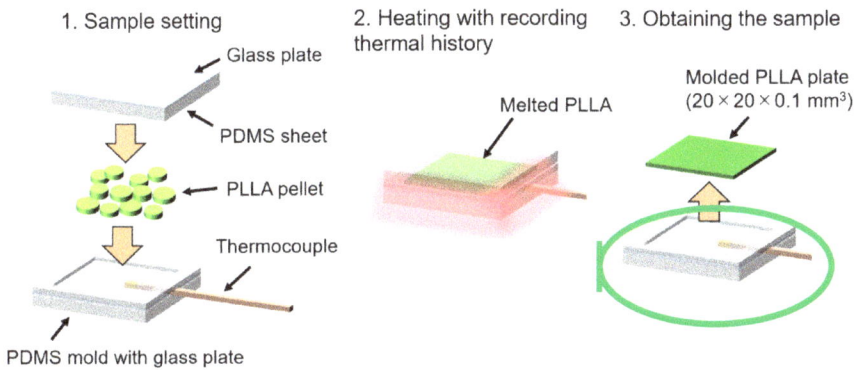

Figure 1. Schematic illustration of the poly (L-lactic acid) (PLLA) micromolding process with recording its thermal history using a poly(dimethylsiloxane) (PDMS) replica mold.

2.3. Microneedle Fabrication Using the Micromolding Process

Figure 2a shows the fabrication process of microneedles using the micromolding process. In the fabrication process of the microneedles with a crystallization time of 0 min, the PLLA pellet was melted into the mold at 220–240 °C, and then, rapidly cooled (>200 °C/min) to room temperature (20–25 °C). In the fabrication process of the PLLA microneedles with crystallization from the molten state (crystallization time $t = 20$ min), the PLLA pellet was melted into the mold at 220–240 °C, and then, the mold was maintained at the crystallization temperature (120–130 °C) for 20 min. Finally, the mold was cooled to room temperature (20–25 °C). A microneedle mold (ST-17, Micropoint Technologies Pte Ltd., Singapore) was used for the fabrication of microneedles with different thermal histories. Figure 2b indicates the microneedle mold dimensions used in the micromolding process.

(a)

1. Sample setting

PLLA pellet

2. Heating for melting and crystallization

3. Obtaining the microneedle

Microneedle mold

(b)

h = 500 μm

L = 200 μm

500 μm

L = 200 μm

L = 200 μm

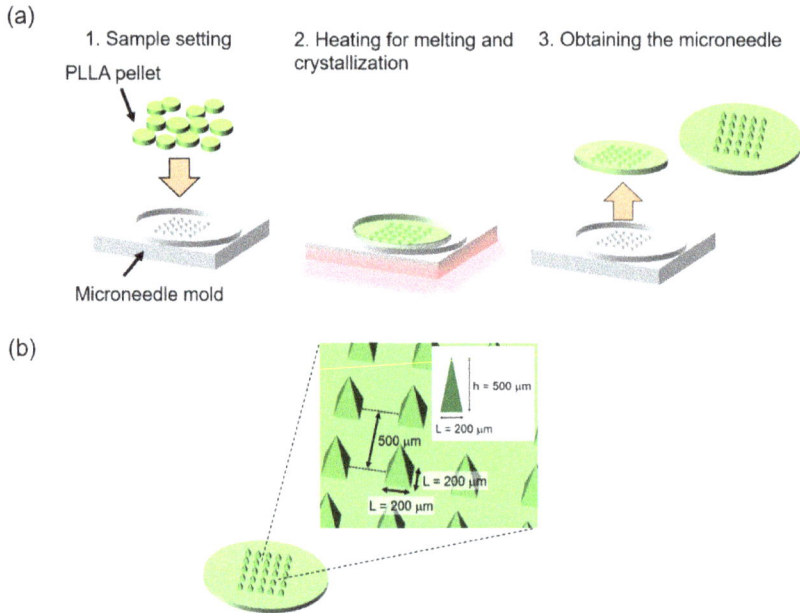

Figure 2. (a) Schematic illustration of the procedures of the micromolding process for fabricating the PLLA microneedles. **(b)** The dimensions of the fabricated PLLA microneedles.

2.4. Characterization

Wide-angle X-ray scattering (WAXS) analysis was performed on the PLLA flat samples using Cu-Kα X-ray sources (λ = 0.154 nm), working at 40 kV and 30 mA, a step size of 0.05 deg, and a scan speed of 20 deg/min (SmartLab, Rigaku Corp., Tokyo, Japan). WAXS analysis was also performed on the PLLA microneedles using the Cu-Kα X-ray sources working at 40 kV and 15 mA, a step size of 0.02 deg, and a scan speed of 10 deg/min (MiniFlex600, Rigaku Corp.). The crystallinity (%) was estimated from $I_c/(I_c + I_a)$, where I_c denotes the diffraction intensity derived from crystalline state and I_a denotes the diffraction intensity derived from the amorphous state from the WAXS spectra, using a software (PDXL, Rigaku Corp.) [17].

3. Results

3.1. Thermal History

The PLLA plate samples were formed using the micromolding process (i) with crystallization from the molten state and (ii) with crystallization from the amorphous state. Figure 3a,b shows the representative thermal histories obtained with processes (i) and (ii), respectively. In process (i), the temperature of the PDMS mold was kept over T_m to form the molten state of PLLA, and then the PDMS mold was kept at crystallization temperature (120–130 °C) and cooled to room temperature. In process (ii), the temperature of the PDMS mold was kept over T_m and then rapidly cooled below T_g to form the amorphous state of PLLA. The rapid cooling rate of >200 °C/min was measured to form the amorphous state in process (ii) as shown in Figure 3b.

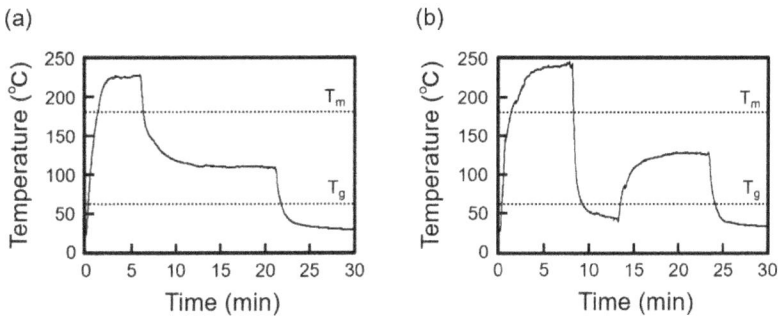

Figure 3. Temperature during the micromolding process. (**a**) After PLLA melting, it was maintained at crystallization temperature (120–130 °C) and then cooled to room temperature (micromolding process (i)). (**b**) After PLLA melting, it was rapidly cooled from >200 °C/min to below T_g and then heated to crystallization temperature (micromolding process (ii)).

3.2. Crystallinity

Figure 4a shows the WAXS pattern of PLLA for the thermal history of process (i). The crystallization times were 2.5, 5.0, 7.5, 10.0, 12.5, 15, and 20 min. Figure 4b shows the WAXS pattern of PLLA for the thermal history of process (ii). The crystallization times were 3.0, 4.0, 5.0, 7.5, 10, and 15 min. The WAXS pattern of PLLA without the crystallization time, which is characterized by a broad band with the maximum value at $2\theta = 16.6°$, indicates a complete amorphous state (Figure 4a,b, indicated as 0 min). The WAXS patterns of PLLA for the thermal histories of processes (i) and (ii) demonstrate sharp peaks at 16.7° and 19.1°, respectively (Figure 4a,b). The intensities of each peak increase depending on the increase of the crystallization time. The peaks at 16.7° and 19.1° were derived from the reflections of 110/200 and 203 planes of the orthorhombic unit cell of the α-form crystal structure of PLLA, respectively [18].

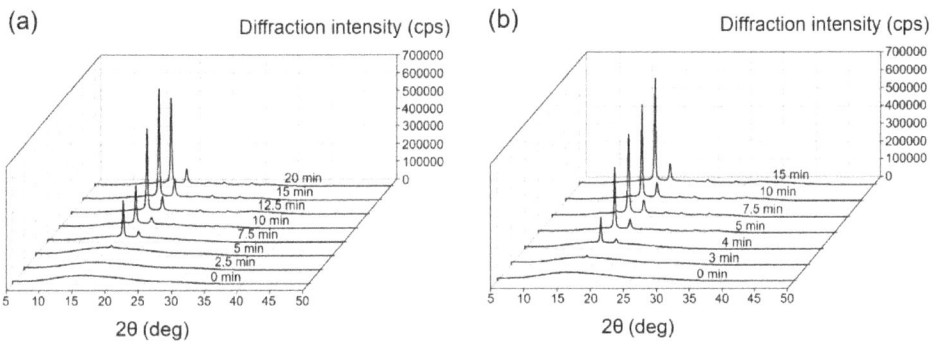

Figure 4. X-ray diffraction peaks obtained from the tested micromolded PLLA samples. (**a**) PLLA formed in micromolding process (i) with crystallization from the molten state. (**b**) PLLA formed in micromolding process (ii) with crystallization from the amorphous state.

3.3. Crystallization Kinetics

Figure 5 shows the degree of the crystallinity (%) of the PLLA that was formed during micromolding processes (i) and (ii). The plots were fitted using the Kolmogorov–Johnson–Mehl–Avrami (KJMA) equation as [19,20]

$$X(t) = X_\infty \left[1 - \exp\left\{ -k(t - \tau)^n \right\} \right],$$ (1)

where $X(t)$ denotes the crystallinity at time t; X_∞ denotes the crystallinity after infinite time; k denotes the overall crystallization rate constant depending on the nucleation and crystal growth rate; n denotes the Avrami exponent; and τ denotes the induction period considered as the period required to form a critical nucleus [21]. Figure 6 shows the plots of $\ln[-\ln(1 - X(t))]$ vs. $\ln(t)$ for the micromolded PLLA substrates during processes (i) and (ii). It is widely known that the inhomogeneous distribution of nuclei results in nonlinearity of the plots, particularly at high-volume crystallinity $X(t)$ [20]. Thus, the curves present a nonlinear end part; however, only the parts used to perform the fitting are shown in Figure 6 [21]. The induction time τ was considered to be 4.0 min for process (i) and 2.5 min for process (ii). The crystallization parameters n and k were 2.3 and 0.011 min^{-n} for process (i) and 2.7 and 0.061 min^{-n} for process (ii), respectively. The n and k values obtained during the micromolding process were consistent with those previously reported for the crystallization of PLLA obtained using differential scanning calorimetry (DSC) data [22–24]. The Avrami exponent value n in the range of 2–3 indicated that mainly a two-dimensional crystal growth was favored [25]. In contrast to process (i), an increase of the overall crystallization rate constant k was obtained during process (ii), accompanied with the quenching operation after PLLA melting. It is considered that the quenching operation formed a large number of crystal nuclei, and thus, the overall crystallization rate constant k increased [26].

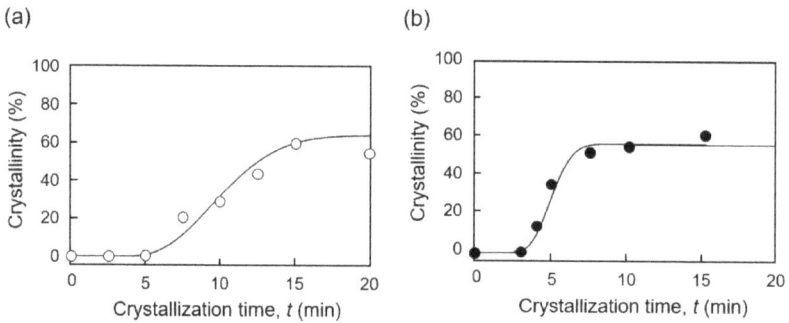

Figure 5. Crystallinity change with time determined using WAXS data. (**a**) PLLA formed in the micromolding process (i) with crystallization from the molten state. Chi-square value $X^2 = 34.7$. (**b**) PLLA formed in the micromolding process (ii) with crystallization from the amorphous state. Chi-square value $X^2 = 20.2$.

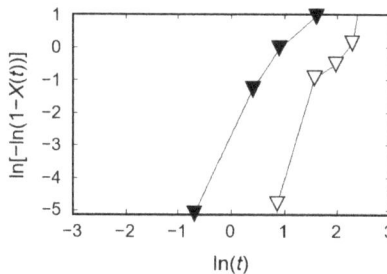

Figure 6. Avrami plots obtained from wide-angle X-ray scattering (WAXS) data: micromolding process (i) with crystallization from the molten state is represented with white triangles, while micromolding process (ii) with crystallization from the amorphous state is represented with black triangles.

3.4. Demonstration of Microneedle Fabrication with Different Thermal History

Finally, we demonstrated the fabrication of microneedles using the micromolding process with different thermal histories. Microneedles are widely used in biomedical microdevices for the drug delivery and vaccination technology [27–34]. The fabrication process of microneedle arrays was performed using the same procedure of forming PLLA sheet samples. PDMS replica molds were used to form microneedles with a needle height of 500 μm and a needle squared base of 200 μm. Figure 7 shows the PLLA microneedle arrays fabricated using the micromolding process with two different thermal histories. The microneedle fabricated with a crystallization time of 0 min was constructed using amorphous PLLA (Figure 7a). WAXS analysis evaluated the crystallinity 0.1%. The PLLA microneedle fabricated with crystallization from the molten state and a crystallization time of 20 min was mainly constructed using crystalline PLLA (Figure 7b). The crystallinity was evaluated as 53.9%. These structural differences of polymer materials due to the differences in the thermal histories of the micromolding process often affect material properties, such as the degradation rate and mechanical strengths. Therefore, the thermal history during the micromolding process should be considered to obtain biomedical microdevices with well-adapted material properties.

(a) (b)

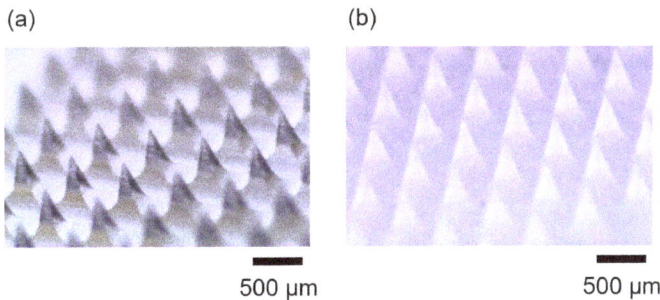

500 μm 500 μm

Figure 7. Photographs of the PLLA microneedle arrays fabricated using micromolding processes with different thermal histories. (**a**) PLLA microneedle array formed during micromolding with a rapid cooling process (>200 °C/min) from the molten state; crystallization time $t = 0$ min. The crystallinity was evaluated as 0.1%. (**b**) PLLA microneedle array formed in the micromolding process with crystallization from the molten state; crystallization time $t = 20$ min. The crystallinity was evaluated as 53.9%.

4. Discussion

Crystallization kinetics is critical when considering the productivity of fabricating products. A slow crystallization rate would result in long processing time in fabricating products. An extremely long processing time would be impractical and economically unfeasible for mass production. Therefore, increasing the crystallization rate in the micromolding process is critical. Controlling the crystallization rate is interesting and several techniques have been investigated by blending specific fillers [15,21]. Such approaches are effective and practical for use in the automotive, electronic, and agriculture sectors as environmentally friendly materials. However, they might be unfeasible in medical applications such as sutures, implants, and microneedles because the safety of the additive materials must be guaranteed. This study showed the crystallization rates of pure PLLA in two main types of crystallization processes during the micromolding fabrication technique, namely crystallization from the molten state, process (i), and crystallization from the amorphous state, process (ii). The crystallization process from the amorphous state (process (ii)), showed a high crystallization rate of 0.061 min^{-n} and might be favorable to shorten the crystallization process time to obtain the fully crystallized products. On the other hand, the crystallization process from the molten state (process (i)) might make it easy to control the crystallinity by controlling the crystallization time and fabricate products with a middle range of about 10%–40% crystallinity.

5. Conclusions

This study investigated the effect of the thermal history during the micromolding process on the crystallinity of a semi-crystalline polymer, namely, PLLA. Two micromolding processes with crystallization from the amorphous and molten states were performed, and their thermal histories were recorded. The crystallinity of PLLA was obtained using WAXS, and crystallization kinetics was analyzed according to the KJMA equation. Compared to the micromolding process with crystallization from the molten state, a crystallization rate of $k = 0.061$ min^{-n} was obtained during the micromolding process with crystallization from the amorphous state, accompanied with the quenching operation forming a large number of crystal nuclei. Finally, PLLA microneedles were fabricated using the micromolding process with two different thermal histories. The thermal history during the micromolding process is important in the fabrication of polymeric microdevices with desired material properties because of its ability to change polymer material morphology.

Author Contributions: Conceptualization, H.T. and T.I.; Investigation, H.T., Y.H. and Y.K.; Writing—original draft, H.T.; Writing—review & editing, Y.H., Y.K. and T.I.; Funding acquisition, H.T. and T.I. All authors have read and agreed to the published version of the manuscript.

Funding: This research was funded by Grants-in-Aid for Scientific Research (grant no. 15K21164 and 19K15416 to HT) from Japan Society for the Promotion of Science of Japan, and the Center of Innovation Program (COI STREAM), from Japan Science and Technology Agency (JST). A part of this research was conducted at Advanced Characterization Nanotechnology Platform of the University of Tokyo, supported by "Nanotechnology Platform" of the Ministry of Education, Culture, Sports, Science and Technology (MEXT), Japan.

Conflicts of Interest: The authors declare no conflict of interest.

References

1. Jamshidian, M.; Tehrany, E.A.; Imran, M.; Jacquot, M.; Desobry, S. Poly-lactic acid: Production, applications, nanocomposites, and release studies. *Compr. Rev. Food Sci. Food Saf.* **2010**, *9*, 552–571. [CrossRef]

2. Södergård, A.; Stolt, M. Properties of lactic acid based polymers and their correlation with composition. *Prog. Polym. Sci.* **2002**, *27*, 1123–1163. [CrossRef]

3. Farah, S.; Anderson, D.G.; Langer, R. Physical and mechanical properties of PLA, and their functions in widespread applications—A comprehensive review. *Adv. Drug Deliv. Rev.* **2016**, *107*, 367–392. [CrossRef] [PubMed]

4. Heckele, M.; Schomburg, W. Review on micro molding of thermoplastic polymers. *J. Micromech. Microeng.* **2003**, *14*, R1. [CrossRef]

5. Zhao, J.; Mayes, R.H.; Chen, G.E.; Xie, H.; Chan, P.S. Effects of process parameters on the micro molding process. *Polym. Eng. Sci.* **2003**, *43*, 1542–1554. [CrossRef]

6. Giboz, J.; Copponnex, T.; Mélé, P. Microinjection molding of thermoplastic polymers: A review. *J. Micromech. Microeng.* **2007**, *17*, R96. [CrossRef]

7. Cam, D.; Hyon, S.-H.; Ikada, Y. Degradation of high molecular weight poly (L-lactide) in alkaline medium. *Biomaterials* **1995**, *16*, 833–843. [CrossRef]

8. Fischer, E.W.; Sterzel, H.J.; Wegner, G. Investigation of the structure of solution grown crystals of lactide copolymers by means of chemical reactions. *Kolloid Z. Z. Polym.* **1973**, *251*, 980–990. [CrossRef]

9. Tsuji, H.; Nakahara, K.; Ikarashi, K. Poly (L-Lactide), 8. High-Temperature Hydrolysis of Poly (L-Lactide) Films with Different Crystallinities and Crystalline Thicknesses in Phosphate-Buffered Solution. *Macromol. Mater. Eng.* **2001**, *286*, 398–406. [CrossRef]

10. Tsuji, H.; Mizuno, A.; Ikada, Y. Properties and morphology of poly (L-lactide). III. Effects of initial crystallinity on long-term in vitro hydrolysis of high molecular weight poly (L-lactide) film in phosphate-buffered solution. *J. Appl. Polym. Sci.* **2000**, *77*, 1452–1464. [CrossRef]

11. Tsuji, H. Hydrolytic Degradation. In *Poly(Lactic Acid)*; Grossman, R.F., Nwabunma, D., Auras, R., Lim, L.-T., Selke, S.E.M., Tsuji, H., Eds.; John Wiley & Sons, Inc.: Hoboken, NJ, USA, 2010; pp. 343–381.

12. Renouf-Glauser, A.C.; Rose, J.; Farrar, D.F.; Cameron, R.E. The effect of crystallinity on the deformation mechanism and bulk mechanical properties of PLLA. *Biomaterials* **2005**, *26*, 5771–5782. [CrossRef] [PubMed]

13. Perego, G.; Cella, G.D.; Bastioli, C. Effect of molecular weight and crystallinity on poly (lactic acid) mechanical properties. *J. Appl. Polym. Sci.* **1996**, *59*, 37–43. [CrossRef]

14. Kanda, Y.; Takehara, H.; Ichiki, T. Mechanical strength evaluation of crystalline poly (L-lactic acid) fabricated by replica micromolding for bioabsorbable microneedle devices. *Jpn. J. Appl. Phys.* **2019**, *58*, SDDK05. [CrossRef]

15. Harris, A.M.; Lee, E.C. Improving mechanical performance of injection molded PLA by controlling crystallinity. *J. Appl. Polym. Sci.* **2008**, *107*, 2246–2255. [CrossRef]

16. Iozzino, V.; Meo, A.D.; Pantani, R. Micromolded polylactic acid with selective degradation rate. *Front. Mater.* **2019**, *6*, 305. [CrossRef]

17. Krepker, M.; Prinz-Setter, O.; Shemesh, R.; Vaxman, A.; Alperstein, D.; Segal, E. Antimicrobial carvacrol-containing polypropylene films: Composition, structure and function. *Polymers* **2018**, *10*, 79. [CrossRef]

18. Zhang, J.; Tashiro, K.; Tsuji, H.; Domb, A.J. Disorder-to-order phase transition and multiple melting behavior of poly (L-lactide) investigated by simultaneous measurements of WAXD and DSC. *Macromolecules* **2008**, *41*, 1352–1357. [CrossRef]

19. Hillier, I. Modified avrami equation for the bulk crystallization kinetics of spherulitic polymers. *J. Polym. Sci. Part A Gen. Pap.* **1965**, *3*, 3067–3078. [CrossRef]

20. Sun, N.; Liu, X.; Lu, K. An explanation to the anomalous Avrami exponent. *Scr. Mater.* **1996**, *34*, 1201–1207. [CrossRef]

21. Battegazzore, D.; Bocchini, S.; Frache, A. Crystallization kinetics of poly (lactic acid)-talc composites. *Express Polym. Lett.* **2011**, *5*, 849–858. [CrossRef]

22. Yu, L.; Liu, H.; Dean, K.; Chen, L. Cold crystallization and postmelting crystallization of PLA plasticized by compressed carbon dioxide. *J. Polym. Sci. Part B Polym. Phys.* **2008**, *46*, 2630–2636. [CrossRef]

23. Castillo, R.V.; Muller, A.J.; Raquez, J.M.; Dubois, P. Crystallization kinetics and morphology of biodegradable double crystalline PLLA-b-PCL diblock copolymers. *Macromolecules* **2010**, *43*, 4149–4160. [CrossRef]

24. Sakai, F.; Nishikawa, K.; Inoue, Y.; Yazawa, K. Nucleation enhancement effect in poly (l-lactide)(PLLA)/poly (ε-caprolactone)(PCL) blend induced by locally activated chain mobility resulting from limited miscibility. *Macromolecules* **2009**, *42*, 8335–8342. [CrossRef]

25. Nagarajan, V.; Mohanty, A.K.; Misra, M. Crystallization behavior and morphology of polylactic acid (PLA) with aromatic sulfonate derivative. *J. Appl. Polym. Sci.* **2016**, *133*. [CrossRef]

26. Mamun, A.; Umemoto, S.; Ishihara, N.; Okui, N. Influence of thermal history on primary nucleation and crystal growth rates of isotactic polystyrene. *Polymer* **2006**, *47*, 5531–5537. [CrossRef]

27. Alkilani, A.Z.; McCrudden, M.T.; Donnelly, R.F. Transdermal drug delivery: Innovative pharmaceutical developments based on disruption of the barrier properties of the stratum corneum. *Pharmaceutics* **2015**, *7*, 438–470. [CrossRef]

28. Tsioris, K.; Raja, W.K.; Pritchard, E.M.; Panilaitis, B.; Kaplan, D.L.; Omenetto, F.G. Fabrication of silk microneedles for controlled-release drug delivery. *Adv. Funct. Mater.* **2012**, *22*, 330–335. [CrossRef]

29. Naito, S.; Ito, Y.; Kiyohara, T.; Kataoka, M.; Ochiai, M.; Takada, K. Antigen-loaded dissolving microneedle array as a novel tool for percutaneous vaccination. *Vaccine* **2012**, *30*, 1191–1197. [CrossRef]

30. Hiraishi, Y.; Nakagawa, T.; Quan, Y.S.; Kamiyama, F.; Hirobe, S.; Okada, N.; Nakagawa, S. Performance and characteristics evaluation of a sodium hyaluronate-based microneedle patch for a transcutaneous drug delivery system. *Int. J. Pharm.* **2013**, *441*, 570–579. [CrossRef]

31. Matsuo, K.; Okamoto, H.; Kawai, Y.; Quan, Y.S.; Kamiyama, F.; Hirobe, S.; Okada, N.; Nakagawa, S. Vaccine efficacy of transcutaneous immunization with amyloid β using a dissolving microneedle array in a mouse model of Alzheimer's disease. *J. Neuroimmunol.* **2014**, *266*, 1–11. [CrossRef]

32. Hirobe, S.; Azukizawa, H.; Hanafusa, T.; Matsuo, K.; Quan, Y.S.; Kamiyama, F.; Katayama, I.; Okada, N.; Nakagawa, S. Clinical study and stability assessment of a novel transcutaneous influenza vaccination using a dissolving microneedle patch. *Biomaterials* **2015**, *57*, 50–58. [CrossRef] [PubMed]

33. Kusamori, K.; Katsumi, H.; Sakai, R.; Hayashi, R.; Hirai, Y.; Tanaka, Y.; Hitomi, K.; Quan, Y.; Kamiyama, F.; Sumida, S.I.; et al. Development of a drug-coated microneedle array and its application for transdermal delivery of interferon alpha. *Biofabrication* **2016**, *8*, 015006. [CrossRef] [PubMed]

34. Mandal, A.; Boopathy, A.V.; Lam, L.K.; Moynihan, K.D.; Welch, M.E.; Bennett, N.R.; Turvey, M.E.; Thai, N.; Love, J.C.; Hammond, P.T.; et al. Cell and fluid sampling microneedle patches for monitoring skin-resident immunity. *Sci. Transl. Med.* **2018**, *10*, eaar2227. [CrossRef] [PubMed]

MDPI

St. Alban-Anlage 66

4052 Basel

Switzerland

Tel. +41 61 683 77 34

Fax +41 61 302 89 18

www.mdpi.com

Micromachines Editorial Office

E-mail: micromachines@mdpi.com

www.mdpi.com/journal/micromachines

MDPI

www.ingramcontent.com/pod-product-compliance
Lightning Source LLC
Chambersburg PA
CBHW080133240326
41458CB00128B/6356